The Technocratic Antarctic

EXPERTISE

CULTURES AND
TECHNOLOGIES
OF KNOWLEDGE

EDITED BY DOMINIC BOYER

A list of titles in this series is available at www.cornellpress.cornell.edu.

THE TECHNOCRATIC ANTARCTIC

An Ethnography of Scientific Expertise and Environmental Governance

JESSICA O'REILLY

CORNELL UNIVERSITY PRESS
ITHACA AND LONDON

First published 2017 by Cornell University Press
First printing, 2017, Cornell Paperbacks

Printed in the United States of America

Library of Congress Cataloging-in-Publication Data

Names: O'Reilly, Jessica, 1978– author.
Title: The technocratic Antarctic : an ethnography of scientific expertise and environmental governance / Jessica O'Reilly.
Description: Ithaca : Cornell University Press, 2017. | Series: Expertise | Includes bibliographical references and index.
Identifiers: LCCN 2016034188 (print) | LCCN 2016035242 (ebook) | ISBN 9780801454127 (cloth : alk. paper) | ISBN 9780801456923 (pbk. : alk. paper) | ISBN 9781501708350 (epub/mobi) | ISBN 9781501708367 (pdf)
Subjects: LCSH: Antarctica—Research. | Scientists—Antarctica. | Ethnology—Antarctica. | Human ecology—Antarctica. | Environmental policy—Antarctica.
Classification: LCC G877 .O74 2017 (print) | LCC G877 (ebook) | DDC 559.89072—dc23
LC record available at https://lccn.loc.gov/2016034188

Cornell University Press strives to use environmentally responsible suppliers and materials to the fullest extent possible in the publishing of its books. Such materials include vegetable-based, low-VOC inks and acid-free papers that are recycled, totally chlorine-free, or partly composed of nonwood fibers. For further information, visit our website at www.cornellpress.cornell.edu.

CONTENTS

Acknowledgments

Working with Antarctic people continues to be engaging, challenging, and fun. So the first debt of gratitude I have is to those in the Antarctic community whom I have worked with over the past ten years, including Jim Barnes, Kees Bastemeijer, Trish Bergmann, Evan Bloom, Julian Chen, Claire Christian, Mark Epstein, Neil Gilbert, Lyn Goldsworthy, Alan Hemmings, Daniela Liggett, Ron Naveen, Polly Penhale, Michelle Rogan-Finnemore, Ricardo Roura, Larry Rudolf, Lou Sanson, Bryan Storey, Tina Tin, Briar Wait, Barry Weeber, and the dozens of people who I interviewed, chatted with, and attended numerous presentations and conferences with over the years. I would particularly like to recognize the members of GCAS 8 for their friendship on the Windless Bight and beyond.

I started with another research plan in place but a long-term scheme to write an ethnography of Antarctica. Thankfully, the faculty and students in the anthropology department at the University of California Santa Cruz veer toward the creative: their collegiality helped turn my scheme into reality. Anna Tsing and Hugh Raffles could not have been better champions or more critical, helpful readers. Don Brenneis, Andrew Mathews, Ravi Rajan, Susan Harding, and Jim Clifford were also tremendous interlocutors while I conducted my research. I would also like to thank Judith Habicht-Mauche for her leadership and encouragement. Fred Deakin, Stephanie Crider, and Debbie Neal provided all sorts of logistical assistance that was sometimes nothing short of miraculous: thank you!

I did part of my writing during my time at University of California San Diego and Princeton University. I would like to acknowledge my mentors Naomi Oreskes and Michael Oppenheimer for their support and help in thinking through climate science from multiple perspectives, as well as the support and engagement of my colleagues Dale Jamieson, Keynyn Brysse, Matthew Shindell, Will Thomas, and Milena Wazeck.

I would like to thank my colleagues at the College of Saint Benedict and Saint John's University: Richard Albares, Ellen Block, Sheila Nelson, Derek Larson, Jean Lavigne, Troy Knight, Matt Lindstrom, Jessica Harkins, Anna Mercedes and many others. I would particularly like to thank Jeff Kamakahi for reading an early version of my entire manuscript. Finally, many thanks to Sheila Hellermann, who helped with many of the logistical requirements of publication with great kindness and professionalism—you are a treasure.

This book has benefited from advice and comments from friends, from roughest draft through the final manuscript. Many provided input over the years, including Jessie Barnes, Eunice Blavascunas, Heath Cabot, Jeremy Campbell, Lindsey Collins, Sandra Koelle, Martha Lampland, Fabiana Li, Ben Orlove, Juno Parrenas, Noah Tamarkin, and Jerome Whitington.

Thank you to the following for financial support of this research: Fulbright-IIE and the wonderful people at the Fulbright New Zealand office, the Social Science Research Council, the Wenner-Gren Foundation for Anthropological Research, the College of Saint Benedict/Saint John's University, the Anthropology Department at the University of California Santa Cruz, and the Graduate Student Association at UCSC.

Portions of chapter 3 were previously published in "Sensing the Ice: Field Science, Models, and Expert Intimacy with Knowledge" in the *Journal of the Royal Anthropological Institute* 22, no. 1, 27–45, in 2016. Portions of chapter 6 were previously published in "Tectonic History and

Gondwanan Geopolitics in the Larsemann Hills, Antarctica" in *PoLAR: Political and Legal Anthropology Review* 34, no. 2, 214–232, in 2011.

I am grateful to work with Cornell University Press to publish this book, particularly the series editor Dominic Boyer, and editors Jim Lance, Susan Specter, and Kate Babbitt. Their insightful and careful comments have vastly improved the finished product—thank you.

My husband, John, is a true partner in this project and in life: thank you for everything. Michael and Quinn, our children, complete our family and bring unimaginable love and joy to our lives. They have been pretty distracting while I've tried to meet writing deadlines, but I'll thank them anyway. Thank you to Grandma, Rachel, Tony, and the O'Reillys. And thank you to my mother and father, whose love, sacrifices, enthusiastic armchair traveling, weather watching, and high expectations have meant the world to me.

Acronyms

ASMA   Antarctic Specially Managed Area
ASOC   Antarctic and Southern Ocean Coalition
ASPA   Antarctic Specially Protected Areas
ATCM   Antarctic Treaty Consultative Meeting
CEP   Committee for Environmental Protection
COMNAP   Council of Managers of National Antarctic Programs
CRAMRA   Convention on the Regulation of Antarctic Mineral Resources
EIA   environmental impact assessment
IAATO   International Association of Antarctic Tour Operators
IGY   International Geophysical Year
IPCC   Intergovernmental Panel on Climate Change
IUCN   International Union for Conservation of Nature
MAF   Ministry of Agriculture and Fisheries
MODIS   Moderate Resolution Imaging *Spectroradiometer*
NGO   nongovernmental organization
NSF   National Science Foundation
PEE   preliminary environmental evaluation
SCAR   Scientific Committee on Antarctic Research

The Technocratic Antarctic

INTRODUCTION

Imagine standing on a crunchy, icy plateau that stretches out to a flat horizon. The sky is overcast and gray and mutes the usual blinding effect of the snow. All around you, the landscape is an expansive gray box mostly devoid of life and color. The Antarctic field camp, a riot of colorful gear, smelly food and fuel, and noisy humans, radically disrupts this scene. Pitched yellow tents, rows of safety flags, and stomped-down footpaths in the snow delineate a human space in the cold, windy monotony. While Antarctica is usually depicted as a place of extreme and overwhelming nature, this book investigates the intensive and contested cultural practices that underpin all human activities on and about the continent.

This book is an ethnographic account of people who call themselves "Antarcticans," the scientists and policy makers who work on Antarctica. In a place with no indigenous people, Antarcticans use expertise as their primary model of governance. Competing claims of nationalism, scientific disciplines, field experiences, and personal relationships among Antarctic environmental managers disrupt the idea of a utopian epistemic community. My analysis focuses on what emerges in Antarctica among the complicated and hybrid forms of science, sociality, politics, and national membership found there.

While I unfold the historical, political, and moral contexts that shape management of the Antarctic environment, I am primarily interested in the technical ways science and policy can work together. Scientists and policy makers often describe these technical means as practices, broadly construed. Scientific research and policy making are practices that inform one another, and the Antarctic environment—with its striking beauty, dramatic human and animal lives, and boring gray days—not only informs science and policy but makes varieties of environmentalism that are shaped by the ways people relate to the place and its inhabitants.

Because the Antarctic is the only continent without indigenous people, Antarctic base employees and scientists form the only social groups in this place. International researchers, support staff, and other visitors do not live permanently on the ice; their stay is limited by time, funding, and challenging conditions. However, Antarctic scientists exert enormous influence in international scientific research and global environmental politics. Their experiences produce information that greatly affects how the Antarctic environment is construed and managed, most importantly with the consideration that some scientists consider Antarctica a pristine laboratory for conducting their research. At the same time, those working on scientific research projects often do not have to comply with the same rules that visitors to the area must follow.

Since the primary Antarctic activity of humans is research, I consider scientists to be not just the arbiters of science but also a community whose members use their expertise to exert authority in environmental management and governance structures. Antarctic scientists are expected to regulate themselves in order to meet environmental standards while working in the field. Technological innovations have made possible some ways for scientists to self-regulate their compliance with environmental standards. In this technocratic milieu, though, scientists strongly influence the continent's regulatory policies, producing techniques that mesh science with policy for the work of managing the environment. In Antarctica, scientists are unusually powerful because of regulations they themselves make.

Antarctica is a seemingly blank space upon which adventurers, explorers, artists, scientists, politicians, animators, and bureaucrats have inscribed their Antarctic visions, their polar world view. Being in Antarctica seems to make people think about their lives and enact particular ways of life for the continent. The place makes and remakes people. Because it is a place without indigenous people, long-term visitors assume the roles of indigenous Antarcticans and governments colonize with panache and without having to contend with the messy problems associated with discovering a place where people have lived for thousands of years. As wild nature, Antarctica is a place where we respond dramatically to the ideas of nature we have learned in our home places—with awe, fear, an urge to protect, a desire to make order, a sense of the sublime.

The Peopling of the Antarctic

Antarctica has a relatively short human history that often seems to start with the International Geophysical Year of 1957–1958. The period before that is deemed the "exploration era," when Amundsen, Scott, Shackleton, Mawson, and others led men to and around Antarctica. While the exploration era is typically contrasted with the peace-and-science era that followed the International Geophysical Year, the contemporary era is always informed by the earlier explorers, in ways that include traditional gender structures and dynamics on the ice, the legacies of observations gleaned in the earliest journeys on the continent, and the primacy and stature implied in being first to a place, planting a flag, and maintaining a territorial claim (Bloom 1993).

The Antarctic governing body is tenuous. It suspends and withholds national territorial claims in favor of maintaining a cooperative transnational scientific continent. In 1959, Antarctica was declared the Antarctic Treaty Territory. Nine nations were the initial signatory members and new nations have been added as they have demonstrated a "scientific presence" in Antarctica. Although nations are on the continent to exercise scientific claims, each one attaches further meanings to Antarctica. Chile has stated that Antarctica is a natural, geological extension of its territory. Norway maintains its claim in order to attempt to standardize whaling in polar seas. Newer member nations, such as Ukraine and India, claim membership among the modernizing, technological, and scientific elite. Member delegates—who are akin to diplomats—meet annually in different nations to discuss and adjust the treaty through unanimous decisions. Foreign service employees, governmental and nongovernmental science advisors, and government lawyers typically form the core of a national Antarctic delegation. Because unanimity is required for resolutions, much Antarctic governance is conducted through informal channels and in good faith. The fact that one nation could effectively and easily end the treaty system if it did not follow such practices sets the Antarctic Treaty apart from international bodies such as the United Nations or the European Union. Antarctic scientists convene their own meetings, discuss and practice environmental regulations in research stations, and advise and oversee their nation's work at Antarctic Treaty consultative meetings. Science and policy are inextricably entangled in Antarctica.

In 1998, the international Antarctic Treaty consultative parties adopted the continent's first comprehensive environmental management guidelines, informally called the Madrid Protocol. This new environmental regime has linked Antarctic governance not only to international cooperation and scientific activities but also to management of the Antarctic continent in terms of a vague but idealistic international environmentalism. In this milieu, a system of governance has emerged in which scientific expertise and authority drives both international geopolitics and the organization and management of the Antarctic environment.

That scientists inhabit the Antarctic is the very justification for the geopolitical institution of the Antarctic Treaty System. The continent is protected in the interest of "peace and science," and almost every piece of Antarctic infrastructure is meant to support scientific research. The current form of Antarctic governance would be unintelligible, almost unimaginable, without scientists and their work.

The practices of Antarctic science are the policies the member nations of the Antarctic Treaty System have created. Scientists inhabit and enliven the decisions made in treaty meetings though their work in the field, although the translation from policy to practice is not simple. Expert embodiment involves the ability to shape the implementation of policy to one's scientific research, which privileges individual scientists and the practicalities of Antarctic field life.

Science and Practice

One rare sunny day shortly before Christmas, my group in Antarctica dug a four-meter pit into the snow to profile the ice sheet. The bottom of the pit, shaded and deep, was −20 degrees Celsius, compared to the relatively comfortable 0 degrees on the surface. I stood at the bottom of the pit while my friend Nadine perched on a carved-out snow ledge above me, writing down the measurements I called out. My job was to figure out the size and shape of the ice crystals at different layers in the profile of the ice shelf. To do this, I pulled out some snow with a tongue depressor that marked each layer. I placed the crystals on a black plastic card with three grids— 1 millimeter, 2 millimeters, 3 millimeters—to determine the general size of the crystals. I called out the size of each crystal to Nadine, who repeated the number back. I repeated the number a third time and then she wrote it down. We had been trained to call out these numbers three times by the climatologist in our camp: it was easy to mishear the called-out numbers in the gusty air and with our mélange of accented English.

After the size of each layer's ice crystals had been recorded, I looked at the shape of each crystal though a loupe and compared it with a card with a grid of drawings of possible snow shapes on it. I compared them and called out a shape code that correlated to the shape that I had determined the crystal resembled (F7, F8, and so on). Nadine confirmed the code number, I repeated it, and she wrote it down.

The meticulous inscriptions scientists mark out over the domains of their research are well documented in science studies. My activities in the bottom of this hole that had been dug over the frozen Antarctic sea reminded me of Latour's (1999) trip to Brazil with a team of soil scientists. His team also relied on low-tech, handheld devices to solidify certainty when comparing little bits of the environment—in their case, a plastic page with holes corresponding to the different shades of brown soils of the Munsell Code. Like the soil scientists, we were trying to make our specific snow pit commensurate with the thousands of other snow profiles and ice crystals that scientists have inspected over time. If we could achieve commensurability, we would have at least the raw data to start to say something about our field site.

This painstaking work, with its careful ritual, marks the first moment when Antarctic policy starts to be performed: initial data collection. Though our hours passed with only a list of numbers to show for it, these data, analyzed and graphed, would later be peer reviewed and published. In the ice pit, we knew that these numbers mattered and we knew why: there is much attention and interest in the fate of the Antarctic ice sheet and the behavior of the ice shelves that buttresses it. Our observations both practiced and conferred expertise. Reading the history of the ice shelf in the way we did not only made us experts (or, in my case, I attempted to be an expert) but also set up a chain of expertise that I expected, or hoped, would flow right through scientific peer review and into science-based policy decisions.

Collecting data accurately is one of the most important practices that scientists can undertake. Data collection matters, and it is imbued with a fetish for accuracy and precision (Wise 1995). Such precision is one reason why scientific work is so difficult to undertake. Antarctic researchers also implement strict environmental practices that can be complex and challenging, particularly under polar field conditions. These environmental practices are heavily informed, if not created by, Antarctic scientists.

The expertise of Antarctic scientists, who have the ability to produce original knowledge about nature and the power this ability affords, drives the governance of Antarctica at all scales, from the micro-edits that scientists do in the field to make policies more workable in the realities of Antarctic fieldwork to the international declarations of efforts related to global problems such as climate change and nonnative species.

In this book, I analyze a cultural group in which everyone is an expert and in which expertise is translated internationally. While other ethnographies have depicted nature and culture as being filtered through the expertise of outsiders, in this case, nature itself is being made and managed by those who study it (Mitchell 2002). I argue that environmental management in Antarctica functions as a technical, knowledge-based bureaucracy with particular historical, scientific, and geopolitical legacies. I use the term *epistemic technocracy* to refer to a mode of governance whose political technologies rely on the form (the discursive habits) and matter (the literal substance) of scientific practice. Simply being from or going to Antarctica makes one a member of the community, and nonhuman nature is included as a participatory member as a matter of course. Knowledge from the physical and life sciences form the basis for the core of regulations that pertain to all human activities in Antarctica. Science is not the "value-added" expertise that legitimizes decisions of governance; it provides the governance framework itself. Moving between science and governance is fraught; although science is explicitly privileged, policy makers must massage scientific expertise to make room for the implicit and explicit expectations of international governance systems.

The international governance of Antarctica's environment is idealized as science-based policy and is carried out as such. This book investigates how this has been made so and the disappointments, successes, and reinterpretations of how governance based on technological expertise—indeed, governance formed and organized as technological expertise—works, when it does. As an epistemic technocracy, the Antarctic provides a model for other locations where scientific expertise and international geopolitics are expected to be in conversation with each other. Experts forge networks with each other, with their audiences, and with their data to technically and politically govern. This changes the conventional concept of technocracy as a form of governance in which skilled experts rule over non-experts to one in which experts govern other experts. While most examples of technocratic governance use skill and expertise as a form of governmental legitimacy, in the Antarctic, technocracy operates through many conversations that involve various forms, modes, and enactments of expertise. The epistemic technocracy model can help dissolve conventional perceptions of the science/policy interface through the enactment of expertise, broadly conceived.

In the Antarctic, people gain expertise through living and working on the continent, but that is not the only way Antarctica confers expertise.[1] The mythical properties of the place, the stories of extremes, exploration, and hardship, are carried forward and given to the people who go there. Antarctic people do not just cultivate and sustain expertise; it is conferred upon them through their lives and work.[2]

While those who work in field sciences produce scientific knowledge, they also grapple with uncertainty, judgment, and public scrutiny of scientific work. This is not new; science has never existed in a vacuum, despite serious attempts to protect science as a domain of objectivity and a sense of integrity, free from politics. Because of the stance that sets scientists apart from the world, science policy often takes the form of "science advice" from experts who then remove themselves from the messy political process (Jasanoff 2005; Latour 2004).

Of course this description of Antarctic governance is idealized—it is part of the emic discourse of Antarctic scientists and policy makers. My analysis focuses on the moments when the ideal model does not work as expected—those mundane and dramatic points when experts must negotiate around, through, and with the ideals they describe. I show that the disruption of an epistemic technocracy, the point when involved experts feel that they are entitled to act and react, reinforces its existence. Mistakes and the work surrounding them help articulate the role of expertise in a form of governance laden with experts.[3]

There are moments when Antarctic policy is based on other things besides science. That is in part because policy and science are not pure categories; they contain epistemic traditions alongside issues such as geopolitical strategies, funding concerns, cross-cultural communication, and nationalism. In general, scientists would like their work to inform policy clearly, but in practice, science (and other forms of expertise) must be refracted through the institutions that make scientific work possible.

Scientists form decisions based on negotiated expertise, as Haas (1990) explains. However, Antarctica's epistemic communities differ from classic geopolitical formulations. Traditional epistemic communities consist of scientific advisory bodies in which scientists generate assessments that they hand over to policy makers, who create policy based on the scientific advice. While the Antarctic Treaty System follows this procedure at the official level, in practice scientists have the power to shape many policy decisions, particularly in terms of compliance.

Some Antarctic policy making is technocratic, though it is rooted in scientific practice. The Antarctic Treaty consultative meetings convene annually in a roughly alphabetical rotation through the treaty member nations. These meetings, like many other international forums, conventions, and agreements, have a front stage and back stage. The work on the front stage—the formal work of the meetings—is polite, translated, deliberate, and procedural. However, much of what gets said on the front stage at the consultative meetings takes shape during backstage deliberations during tea breaks, at lunches and receptions, in informal working groups, and by the lawyers who write the binding language. Papers are written and circulated beforehand, and intersessional work, usually conducted through online forums, attempts to carry forward the objectives of a particular Antarctic Treaty consultative meeting. Some Antarctic policy makers are diplomats with long-term positions at their country's "Antarctic desk," and some are part of a revolving cadre of professionals doing a stint of just a few years. A mix of technocratic strategies and personal relationships shape Antarctic policy making.

International technocrats move policy through the Antarctic Treaty System, and it is important to take technocracy—the combination of skill, expertise, and bureaucracy—seriously. Bureaucracy is not a place where good ideas go to die: things happen, they come to life in particular ways in bureaucracies. Bureaucracy and technocracy are generative, and the people within them are active, creative, and participatory. Working with, against, or alongside such organizational structures helps us understand how people collaborate within them, even when they are acting in ways that seem counter to the formal goals of an organization (Hull 2012; Strathern 2000; Weber 1947).

Antarctica presents us with an opportunity to study a "culture of expertise" in a place where there assumed to be no other culture. Moreover, as a community of scientists, Antarctic people have and exercise a unique opportunity to fuse together scientific expertise and governance—to make policy decisions that emerge directly from scientific knowledge. Elsewhere, this idea is a technocratic chimera, something that is impossible because of plurality and multilayered and overlapping jurisdictions. The idea that scientific knowledge can directly inform policy might also be unobtainable in Antarctica, but it at least holds the potential of being possible. This ethnography follows the people who attempt to make scientific knowledge and environmental policy relevant to each other in idealistic, partial, and hopeful ways.

Methods and Field Sites

The research for this book was conducted primarily in New Zealand, home of the "Antarctic Gateway" city of Christchurch, and on an expedition to Windless Bight, Antarctica, with the New Zealand Antarctic Program. But this project also accompanies the transnational journeys Antarctic scientists and policy makers make to temporarily styled Antarctic places such as science conferences, policy workshops, and the international Antarctic Treaty consultative meetings in Scotland, Australia, and India. I examine the lived intricacies of this international environmental space and people's relationships to Antarctic environmental management by mapping, examining, and traveling within the networks scientists and other Antarctic community members form. Because competing claims of nationalism, scientific disciplines, field experiences, and personal relationships in Antarctic environmental management disrupt the idea of a utopian epistemic community, I focus on what emerges in Antarctica among the complicated and hybrid forms of science, sociality, politics, and national membership found there.

My primary fieldwork lasted fourteen months in 2005–2006 and took place in Antarctica, New Zealand, Australia, and Scotland. I also conducted some introductory research in Christchurch, New Zealand, in December 2003 and some follow-up research at the Antarctic Treaty Consultative Meeting in New Delhi, India, in 2007. My involvement with the treaty system is ongoing.

The methods I use are traditional ethnographic participant observation and interviews in Antarctica, in labs, at policy meetings, and at scientific conferences. I participated in an Antarctic field expedition that conducted environmental monitoring projects on the ice shelf, on a group of Weddell Seals, on erosion in ice-free areas, and a weather monitoring program. Then I tracked the data we gathered and those collected by other Antarctic scientists through the processes of local, national, and international Antarctic environmental policy making. My approach to writing focuses on case studies of instances when epistemic technocracy fails to run smoothly; my ethnography is one of hiccups, slip-ups, and problems of various scales. In these problematic case studies—of samples left in airport freezers, of scientists facing a public skeptical about global warming, of governmental regulations that require extreme levels of care regarding Antarctic rocks—members of the epistemic technocracy use their expertise to solve problems. It is these moments, large and small, when Antarctic scientists and policy makers form the international management of Antarctica.

My fieldwork was divided into three major periods of activity. For the first four months, I trained for life and work in an Antarctic field camp with the members of an Antarctic research expedition. (For two months following my return from the Antarctic, I continued working with this group, preparing scientific and environmental reports and presenting the group's findings.) For the second set of four months, I was intensely involved with local preparations for the Antarctic Treaty Consultative Meeting that was to take place in 2007. Workshops, policy drafts, and meetings dominated this period of time, as New Zealand scientists and policy makers negotiated the policy views that would be articulated at the meetings. For the remaining time, I tracked in more detail several of the most compelling cases that had emerged earlier during my research. Throughout my research period, I conducted several series of ethnographic interviews, including with several sets of Antarctic scientists, with Antarctic scientists who have become policy makers, and with Antarctic policy makers. I attended and recorded twenty-four public presentations by Antarctic policy makers and scientists and observed several Antarctic science and policy conferences, meetings, and workshops.

I followed Antarctic policy makers and scientists as they conducted their work. In Antarctica, I stayed three nights at New Zealand's Scott Base and spent two weeks in an Antarctic field camp, where I conducted scientific research activities and learned to live in Antarctic conditions. I spent most of my fieldwork time in New Zealand, where I was based in the South Island city of Christchurch.[4] The city is a primary academic center, tourist destination, and logistical base for several national Antarctic programs. While in other parts of the world the Antarctic is a dream, a distant mirage, in Christchurch, Antarctica informs everyday public life. A statue of the ill-fated Antarctic expeditioner Robert Falcon Scott once stood by the Avon River near the Cathedral Square downtown. The Canterbury Museum boasts a large Antarctic display, and during the time I lived a few blocks from the museum, it set up the decommissioned Antarctic Hallett Station for permanent display in one of its halls. Bailies Bar on the square is a traditional site for pre-departure dinners and drinks.

The imposing and modern-looking International Antarctic Centre is a few miles out of town, near Christchurch International Airport. On the campus of the centre are the main offices for Antarctica New Zealand, the national Antarctic program. The centre also houses the offices of the national Antarctic program of the United States, which is run by the National Science Foundation, and the Italian national Antarctic program. Warehouses full of gear and supplies lie on the outskirts of the property, as does a branch of the United States Postal Service. People flying south check in at a designated terminal at the International Antarctic Centre, wait in the waiting room, watch pre-departure videos, load onto a bus, are driven to the massive U.S. Air Force DC-17s or the anachronistic and hardy New Zealand Royal Air Force Hercules, and take off for the continent.

Next door to all of these official governmental Antarctic activities is the Antarctic Attraction, a privately owned, somewhat expensive tourist facility that has displays on many Antarctic topics: Scott Base, flora and fauna, field research life, and, of course, penguins. One room, named The Antarctic Storm, is full of ice and snow. Visitors don bright red parkas before entering, and every half hour or so, a "storm" begins in the room as wind kicks up and the temperature drops. This high-concept, experiential museum has been extremely successful and lucrative since it opened and further consolidates Christchurch's reputation as an Antarctic gateway community.

For the duration of my research, I found an institutional home at an interdisciplinary university department at the University of Canterbury called Gateway Antarctica. The department is small; it consists of a geologist, an Antarctic policy specialist, a geographer, and a handful of PhD students and staff, but it makes significant contributions to consolidating Antarctic expertise in the area. Many faculty members from other University of Canterbury departments and other universities are allied with Gateway Antarctica. The department also organizes many public lectures, both on campus and in downtown Christchurch, often capitalizing on visits from international Antarctic scientists stopping over in Christchurch on their way to or from Antarctica. Gateway Antarctica also organizes a summer program for graduate students called the Graduate Certificate in Antarctic Studies, and it was through this program that I was able to travel to the Antarctic.

The international Antarctic Treaty Consultative Meeting (ATCM)—basically, a series of international diplomatic meetings about the continent—convenes annually in one of the signatory countries. During my research, the 2006 ATCM was organized by the United Kingdom in Edinburgh, Scotland. I attended these two-week-long meetings as a delegate for the Antarctic and Southern Ocean Coalition (ASOC), an environmental NGO that has expert status at the meetings. In 2007, I attended the

ATCM with ASOC in New Delhi, India. The temporary but significant presence of an international cadre of Antarctic policy makers and experts in a country often sparks public events, local media attention, and Antarctic science lectures and artistic and historical displays in the city where the ATCM are being held. For two weeks or so, Antarctica emerges in the world's cities.

As I was based in New Zealand for most of my fieldwork, much of my findings are heavily influenced by my experiences in this country with New Zealand Antarctic scientists, activists, managers, and policy makers. As I described above, Christchurch is one of several Antarctic gateway cities on the planet. While this can easily be attributed to the city's geographic proximity to the Antarctic, the fact that the city of Christchurch and the nation of New Zealand are wrapped up in Antarctic matters stems from their history of British colonization and progressive politics.[5] There is a keen sense of Antarctic heritage among involved New Zealanders.

Overview of the Book

This ethnography analyzes the ways Antarctic science, policy, and governance complicate each other. These complications emerge as scientists, policy makers, Antarctic life forms and objects, data, and paperwork are arranged to influence environmental management on the continent. Each of the following chapters presents a case study that analyzes the technical ways that science and policy merge in Antarctic environmental management.

Chapter 1 opens with a history of how Antarctica has been imagined through the year, even before humans found it. How do these imaginations continue to inform our perceptions and policies for Antarctica in contemporary times? By exploring how history entwines with visions of the future, this chapter sets the stage for analyzing expert understandings of the governance needs of Antarctica. In chapter 2, I take readers into the recent historical context (1959–2006) of environmental management and policy in the Antarctic. I focus on a very specific history of an environmental micropractice that is mandated in the McMurdo Dry Valleys: moving Antarctic rocks back to their original location. This case study shows how the current global environmental regime of the Madrid Protocol reconfigures people's technical and material relationships with the Antarctic environment. In chapter 3, I analyze how experts sense, interact with, and read expert knowledge from data (in this case, about ice). Sensory engagement among scientists is not primitive, elemental, basic, or instinctual but is instead bound up in the complexities of nationalism, scientific translations of scale, and boundary skirmishes over what counts as expertise within scientific disciplines. Expertise is formed through intimate encounters and cultural learning that teaches experts in the making how to relate, analyze, compare, and interpret. In chapter 4, I describe the process of making new environmental policies for Antarctica with the example of nonnative species, an emergent problem turned policy. This chapter focuses on how policies related to nonnative species create challenges for the scientists who must follow the policies when they carry their samples and specimens out of Antarctica. In chapter 5, I use the issue of biosecurity to illuminate how technocrats move Antarctic field science–based policy into international diplomacy. In chapter 6, we visit the Larsemann Hills, site of both a proposed Antarctic Specially Managed Area and a new Indian research station. Here, complex articulations of nations' "right to develop" rub up against claims of repetitive science and a "failure" of India to "understand the process," providing a case in which expertise is contested, particularly in terms of scientific research, in international governance. In chapter 7, I examine the role of Antarctic scientists and their data at presentations about climate change to public audiences. Using certain prominent examples of climate changes in the Antarctic, scientists situate Antarctica as a place of climate change, establish themselves as expert spokespeople for climate change, and craft complicated political messages in their presentations. The political work Antarcticans do must be presented in an environment that is far removed from the genteel deference of Antarctic Treaty System policy meetings and into a world of less stable deference to scientific and policy expertise.

The chapters that follow are case studies that depict how knowledge-based communities form and have effects on the world. Doing so requires work that involves at least tentative agreements about the core ideas of expertise and communities. Antarctic people translate science through the policy system in order to make decisions about how to manage the environment through formal and informal arrangements of policy and practice, of humans and nonhuman entities. Policy and practice in these Antarctic expert communities coalesce as procedures, documents, and audiences are constructed. In the lives and work of members of the community of experts, this continent of peace, science, and other exceptionalisms is crafted as an epistemic technocracy.

1

The Imagined Antarctic

How does time spent traveling to, living in, and working for Antarctica shape the way the place is imagined, not only by seasoned Antarctic policy makers, base workers, and scientists but also by those people who will never visit the ice? Environmental imaginaries shape what kinds of environmental policy and management are possible. The Antarctic is framed by a transnational policy climate that includes an expansive menu of management options to choose from, idealistic managers to implement these options, and a place that inspires some to take Herculean steps to protect it. The ways people have imagined Antarctica over time provoke understandings of the continent as a site of exceptions that emerge from an often-repeated statement that it is the highest, windiest, driest, and coldest continent. It is these extremes that have rendered Antarctica the largest remaining swath of unpopulated space on Earth—an exception to widespread human development. It is the only large landmass on Earth not represented by the United Nations. This exceptionalism is the reason for constant negotiations about whether Antarctica is a nationless, international space or whether it is, in fact, its own entity. The ways Antarctica is produced as an imagined environment—as an entrance to another world within our world; a home to unknown, potentially dangerous species; an endless, foreboding expanse of unnavigable ice; a site where nationalism and empire will expand or end; a test of humanity's character and fortitude—are carried forward in present-day habitations of and meanings of the continent.[1] This chapter considers imaginations of what Antarctic was before humans discovered it and the imaginations of explorers, scientists, base employees, and policy makers about the continent. These imaginations haunt the Antarctic, particularly how it is seen as exceptional and the ways this exceptionalism is both upheld and contested by quotidian Antarctic life.

Coming to Know the Southern Unknown Land

The long imaginings about Antarctica, spanning continents, cultures, and historical moments, is integral to the ways it continues to be imagined and depicted today. Imagined Antarcticas are also rooted in the histories of scientific knowledge and exploration. While I could compare imagined, pre-discovery Antarctica with today's "real" continent, I take as foundational the notion that imaginings about the Antarctic environment continue to be shaped by histories of science, discovery, exploration, geopolitics, environmentalism, and wonderings about faraway places. The imagined environment is not just a social construction; it is also nature impacted by human action and decisions.

Greeks and Polynesians imagined something like Antarctica millennia before the continent was discovered. Around 530 BC, Greek Pythagoreans postulated that the earth was round. Because of a concern with balance and symmetry, this round earth hypothesis carried a suggestion that in order to balance all of the known landmasses in the Northern Hemisphere, there had to be similarly weighty landmasses on the southern half of the earth. The Pythagoreans suggested such a continent and named it *Antarktos*, meaning "opposite the Arctic." Later, exploration-era maps that predate sightings of the continent by Westerners included the Terra Australis incognita, or the southern unknown land on the bottom of the planet, an extension of the Pythagorean idea of Antarctica.

Figure 1. Descriptio terræ subaustralis. Petrus Bertius, Descriptio terræ subaustralis (Amsterdam: J. Hondius, 1616), Historic Maps Collection, Department of Rare Books and Special Collections, Princeton University Library.

Pacific Islanders tell a story about a skilled navigator who piloted his canoe south until the sea turned completely to ice. In some recollections, the traveler is Rarotongan (a Cook Islander) and in others, he is Maori (a native New Zealander). Named Ui-te-Rangiora, he navigated south to limitless sea ice around AD 650. The legacy of his exploration spread throughout the Polynesian Islands, and when James Cook reached the tropics, most of the people he encountered knew there was a great icy expanse awaiting him as he traveled further south. The combination of the technologies of Polynesian navigation by the stars and shared expertise among the region's island inhabitants produced knowledge of what to expect when traveling south.

In Western exploration, the ability to measure, record, and announce one's cartographic achievements using standardized measuring procedures was of the utmost importance. Cook's expedition was the first to certifiably cross the Antarctic Circle, though Ui-te-Rangiora may have preceded him by over a millennium. In the southern summer of 1772–1773, when

Cook crossed the Antarctic Circle, he had the expertise to record the event for posterity and managed to circumnavigate the Antarctic continent even though he never laid eyes on it (Thomas 2003). That season has been estimated to have been particularly cold, so perhaps the ice pack near the continent's coast did not break up as usual, keeping Cook and his ships from spotting the continent. In any case, Cook determined that he had traveled sufficiently far south and had seen enough endless ice to know that no one else would ever need to repeat such an endeavor. In his journal, Cook wrote on February 6, 1775, "I firmly believe there is a tract of land near the Pole" because of the large quantities of ice he encountered in the Southern Ocean. Cook (1961) continued his journal entry by claiming that this land had to be inaccessible because of the ice and wrote that "the risk one runs in exploring a coast in these unknown and Icy seas, is so very great, that I can be bold to say that no man will ever venture farther that I have done and that the lands which may lie to the South will never be explored" (637–638). Present-day Antarctic scientists like this quote, sometimes citing it at the beginning of their presentations. Perhaps the irony appeals to their ideas about working in a frontier zone or about reproducing and exceeding earlier attempts to explore the Antarctic.

Dozens of other European traders and explorers traveled the Southern Ocean contemporaneously with Cook, making the first known sightings of tabular icebergs and discovering many of the sub-Antarctic islands. A Russian expedition south that was captained by Thaddeus von Bellingshausen in 1820 was the first that is known to have sighted the continent. Although seal populations all over the sub-Antarctic islands collapsed in the late nineteenth century, the sealing industry and its impacts on the environment continue to shape Antarctic politics today. Captain John Davis, who was on a sealing expedition in 1821, is the first person who is known to have set foot on the Antarctic continent.

Landing on Antarctica did not stop dreamers from constructing beautiful and fanciful imaginaries of the continent. One sculptor of imagined Antarctic environments is John Symmes, an American businessman who encouraged the United States Congress to fund Arctic and Antarctic expeditions in the early 1800s. Symmes pursued a scientific mission that compelled him to reach out to the public and to the government to support his concern. His ideas were based on theories about the earth that were first suggested by the astronomer Edmund Halley in 1692. Halley believed the earth was hollow and that its interior was populated with life. The holes that led to the center of the earth were to be found at the North and South Poles. He suggested that the aurora borealis was the result of gasses escaping from the hollow earth. Symmes was one of several who suggested that an advanced civilization of beings lived within the seven nestled spheres inside the hollow earth.

Symmes constructed maps and globes depicting the world with an entrance at both poles to the hollow earth, postulated how large each of these entrances was (4,000 miles wide at the North Pole and 6,000 miles wide at the South Pole), and presented his theories to well-attended, though widely giggled at, public lectures. Symmes was often described as insane. In a move befitting a scientist who knew how to insert himself into political and bureaucratic systems, he had himself examined by a psychiatrist and certified as sane.[2] He made this certification available to the public as he continued his educational tours (Fleming 2001; McCaughrean 2007; Wilson 2003).

Congress funded an expedition to the Arctic to see if Symmes's hole existed. Those explorers did not find it. Symmes did not live to witness the fact that his theories helped convince Congress to fund the Wilkes Expedition in 1838, the first U.S. government–funded expedition to Antarctica. One of Symmes's proponents, Jeremiah Reynolds, was asked to participate so that he could help the party look for a giant southern opening in the earth.

While the existence of Symmes's hole was eventually disproven, the fact that Symmes rooted his work in science encouraged exploration of the North and South Poles. Symmes deserves credit as the inspiration for fiction that includes Edgar Allen Poe's *The Narrative of Arthur Gordon Pym* (1838), Jules Verne's *Journey to the Center of the Earth* (1864), and Edgar Rice Burroughs's *Pellucidar* series (1922, 1923, 1930a, 1930b, 1937, 1944, 1963). His legacy is also apparent in science fiction films set in the Antarctic—particularly *The Thing* (1982), based on the 1938 short story "Who Goes There?" by John W. Campbell Jr.—in which research station employees violently encounter a population of nonhuman creatures.

Some "alternative scientists" still believe that Symmes's hole exists: stories of ships disappearing while traveling to the north, peculiarities in satellite photography, and other technologies and stories keep the theory simmering. In 1906, a man named William Reed published *The Phantom*

of the Poles, which asserted that the poles could never actually be discovered because they do not exist except as holes to the earth's interior. These
conspiratorial narratives were echoed in the Cold War–era space race, a
period that also inspired the modes of Antarctic exceptionalism that continue to be apparent today.

As sealers, whalers, and government-commissioned expeditioners landed
more frequently on various points of the Antarctic coastline, explorers
began organizing attempts to reach the South Pole. The key figures of this
"race to the pole" were Robert Falcon Scott, Roald Amundsen, and Er-
nest Shackleton. Scott and Amundsen's teams were on the ice at the same
time—1911—with drastically different outcomes. Before I lived in New
Zealand, I had heard the story of Scott's party as one of the last gasps of the
dying British Empire. The fate of its members marked the end of an era,
making space for the United States to achieve superpower status. In New
Zealand, however, and I suspect elsewhere in the former British Empire,
Scott continues to be alternately critiqued or hailed as a hero who proved

Figure 2. The Hollow Earth. Reed 1906.

the strength and grace inherent among the British people, even when facing an icy death.

Scott's Terra Nova Expedition made it to the pole, but when they got there they learned that the Norwegians, led by Amundsen, had reached it a month earlier. (Amundsen figured only as a passing character in the exploration history narratives I heard in New Zealand.) Scott's group had pressed on to the pole even though they knew that they did not have enough supplies for the return journey. One expedition member, Lawrence Oates, famously walked away from the camp on the return trip into a snowstorm, remarking that "I'm just going outside and I may be some time." He never returned. The interpretation is usually that Oates sacrificed himself for the good of the rest of the Terra Nova Expedition. Unfortunately, the entire party perished on the return journey. In his final journal entry, Scott wrote, "I do not regret this journey, which has shown that Englishmen can endure hardships, help one another, and meet death with as great a fortitude as ever in the past."

The well-known story of the demise of the Scott party haunts the Antarctic places and people I encountered in my fieldwork. The Terra Nova Expedition comes up in both utterly serious and completely joking ways. The Scott statue in figure 3 was located in the city center of Christchurch, New Zealand, about three blocks from the flat I lived in while conducting my research. After it sustained damage in the Christchurch earthquake,

Figure 3. Scott Statue in Christchurch. Photo by author.

it was relocated. Scott's famous quote is inscribed in the granite below his figure. Memorial services for Scott and more recent deaths of people involved in Antarctic affairs occur at this statue yearly.

The group I traveled to Antarctica with carried strands of the Scott expedition's story. One member of our group had a tendency to fall asleep during our pre-departure meetings and classes. When we would go to pubs or out to dinner as a group in the evenings, he would invariably disappear for up to an hour, returning to our table with no comment. Our group speculated about where he went and what he did during these times. One person began to call him Oates because he would always "be some time." Our group's Oates was dismissed from our program before we went to Antarctica.

I couldn't shake the specter of Scott's party during my first night camped on the ice shelf at Windless Bight, Antarctica. We had already been in Antarctica a few days, staying in bunk rooms in New Zealand's Scott Base. I had recently learned while the belongings, photographs, journals, and letters of the members of the Scott Party had been retrieved a year after their deaths, their bodies had been left in their tents on the ice shelf. Someone had calculated how deep the snow covering the party now was and had estimated when their bodies would eventually leave the continent, encased in miles of ice as an iceberg calved off the mainland.

As I lay in the bizarre yellow light of my tent, trying to fall asleep in the persistent sunlight, I could not stop thinking about the members of the Terra Nova Expedition and their frozen bodies slowly migrating with the flow of the continent's ice. Had they frozen to death or starved to death? I assessed my own situation. I certainly wouldn't starve to death, as I had a huge duffel bag full of food in the vestibule, about two feet from my head; it contained canned fruit, instant noodles, chocolate bars, Marmite, and Christmas puddings in plastic containers. But would I die if I somehow fell asleep, managed to get myself out of my sleeping bag's three lofty layers, and somehow rolled off my thick inflatable mattress to sleep on the snow? On that first night on the ice, the thought that I would wake up if I ended up lying unprotected on snow never occurred to me.

Shackleton is alternately portrayed by biographers as a bumbling failure or as the mastermind of one of the greatest survival stories of all time. People love the way he is supposed to have recruited participants for the Endurance Expedition (1914–1916) through an ad in a London newspaper: "Men Wanted for Hazardous Journey, Small Wages, Bitter Cold, Long Months of Complete Darkness, Constant Danger, Safe Return Doubtful. Honor and Recognition in Case of Success." The source of the quote has not been verified, though it is often cited.[3] Apparently this description was adequate enticement. After Shackleton failed in his attempt to be the first to reach the South Pole, he intended this journey to be the one where the first group of men completed a trek across the Antarctic. This did not happen either, as the ship *The Endurance* became locked in the pack ice about twenty miles offshore. The crew wintered in the ship until the pack ice cracked up their vessel one evening. Men and dogs slid out of the boat onto their lifeboats and set up camp on the pack ice until it broke up. An 800-mile open-water voyage in one of these boats followed by a 150-mile trek through the unexplored, mountainous interior of South Georgia Island to a whaling camp led to the rescue of every single seaman after twenty-two months.

In a PBS documentary about the expedition, the survival of every man in the expedition was juxtaposed with the lives and deaths of several species of other animals.[4] As in all Antarctic stories, animals and the experiences of humans with them are woven prominently into the narrative. The expedition's photographer loved to capture the images and the antics of the several teams of huskies who sailed with them. The documentary cuts to images of a dog's breath condensing in thick clouds as *The Endurance* becomes locked in the ice. The crew spent their time icebound training the sled dogs to obey various commands, a process that was duly filmed, and several journal entries indicate that the most exciting event of that long, dark winter was the birth of four husky puppies onboard. The documentarians spent plenty of time featuring the scenes of dogs working and entertaining the men on the ship. In the PBS program, the turning point of the first happy but dire months during the first winter on *The Endurance* occurs when the men and dogs have evacuated the ship, supplies are dwindling, and they kill first the puppies, who will not be able to survive the conditions, and then the teams, since they are consuming resources. The ice breaks up, and the men without dogs conduct a harrowing five-day row to Elephant Island.

The men, we are told, are mostly entirely insane as they row without sleep or food for days on end. Shackleton and his officers worry in their journals that their subordinates are mutinous and/or will die as they row through the night. Gloriously, in the morning they spot Elephant Island, row to shore, and do the following: 1) embark on a massive spree during which they kill many more seals than they need for food; a journal entry described it as "slaughter for slaughter's sake—a behavior which, in many cases, could be described merely as insane"; 2) eat their first hot meal in five days; and 3) sleep. As Shackleton and a few select crew members set off for South Georgia Island, the remainder of the crew waits for rescue on Elephant Island, almost starving as they eat the island's penguins and seals to extinction. Despite arduous effort (this trip was one of Shackleton's four attempts to cross the continent on foot), Shackleton and his crew never reached the South Pole.

The hauntings of people and animals who have previously traveled to Antarctica help us imagine Antarctica as a place of struggle, extreme conditions, hazard, sacrifice, challenge, adventure, and death. Antarctic exploration narratives shift over time as the writers position themselves and Antarctica differently in relation to global geopolitics, environmental ethics, and colonial sensibilities. Exploration is an activity that inspires and requires the positioning of people and ships in relation to the seas, to other nations, and to other explorers. This positioning work was done by Antarctic explorers, and the narratives about them, were and are explicitly bound up with the colonial enterprises of the British Empire, Western science, and masculine adventure. Later, advocates for the environment pointed to the actions of these early explorers to support arguments that it was imperative that humans interact with Antarctica in different ways.

Nature exploration and scientific discovery have been at the forefront of European travels since the eighteenth century. That is when scientists achieved the ability to claim sites as resources and manage them, forming early iterations of environmental governance (Carter 1987; Miller and Reill 1996; Pratt 1992). The long history of science and colonialism, particularly among the British colonies, continues to play out (and be disputed, critiqued, and rebelled against) in contemporary Antarctic science and policy, particularly in international Antarctic geopolitics.

Although Antarctic exploration is colonial in the sense that it carries overtones of conquering the world, it lacks the colonial work of reconfiguring social relationships toward the metropole (Byrd 1930; Cherry-Garrard 1929; Cook 1900). Bloom (1993), who argues that the aura of heroic masculinity is connected to visions of national expansion, considers polar exploration to be a final filling in of blank spaces on a map. But because commercial possibilities in the Arctic and Antarctic were few in the eighteenth century, expansionists turned to science as a justification for exploring these areas. Bloom writes of the North Pole, though her statement easily applies to the south as well:

> As pointless as a trek across a barren wasteland may have seemed to those concerned with financial gain, such an exploit has a pervasive scientific appeal. It literalized the colonial fantasy of a tabula rasa where people, history, and culture vanish. The absence of land, peoples, or wildlife to conquer gave polar exploration an aesthetic dimension that allowed the discovery of the North Pole to appear above political and commercial concerns. Thus, paradoxically, it was the lack of material gain from such an exploit that transformed polar exploration into a new kind of imperial theater with all its colonial and scientific trappings. (2–3)

The exceptional blankness of the polar region, its role as *terra nullius*, purified it into an easily imagined scientific place where little to no indigenous meaning could complicate things.

Historians of exploration have often written their characters in a way that ascribes certain genders, nationalities, and dispositions to them. Gurney's *Below the Convergence* (1997), for example, describes the humans in and around Antarctica during the exploration era. In Gurney's telling, explorers were the predecessors of the type of person who was expected to travel to or live in Antarctica in a later time, including the present. While the general theme of Gurney's book is accomplishing adventurous feats in extreme conditions (weather, mutinies, scurvy), he weaves other dominant threads into his narrative. One thread is the stories of boyish antics that complement the grave peril these explorers faced. Men toast their wives and sweethearts ("may the two never meet"), carouse onshore, and fire cannons at icebergs to watch them crumble. Another thread is the occasional commentary on present circumstances, couched in moral terms (for example, the presence of a fourteen-year-old ship's boy sets off a brief statement bemoaning today's "extended adolescence"). Similarly, Gurney compared the "Yankees" and the British: for example, "Americans take their bathrooms when they travel, the British take their pubs" (240). Finally, and less explicitly, nature is written into these accounts—violent, frightened, smelly, wet, cold, impossible nature—that incessantly, determinedly permeates the "sterile stretch" that southern explorers see.

Accounts of more recent Antarctic exploration have continued the tradition of depicting the Antarctic as a place that cements nationalist claims and national character, building on the ideology of sagas of the first exploratory expeditions in order to create an ideology of nationalistic logic in the Antarctic (Fuchs 1982; Templeton 2000). The Antarctic is a site where geopolitics, nationalism, global science, and environmentalism flourish. This is an example of metalepsis, "invocations of history that are deployed in order to authorize even as they are stripped of their authority and content" (Burnett 2000, 39). Metalepsis allows generations of explorers to build authority and validation for themselves through referring to previous exploration legacies. I find a similar phenomenon in more recent firsthand adventure narratives about Antarctica. These stories are usually more narrowly focused in terms of geography or topic, and they usually offer overt tales of preservation instead of those of adventure and hardship (Brown 1991; Green and Potten 2003). In such narratives, Antarctica becomes a place of political maneuvers in the name of geopolitics and the environment, and explorers use their journeys and writing to compare Antarctica with the rest of the world. Antarctic exploration narratives value the environment in various ways that can be described as colonial exploitation: it is not just exploration but other values that tie science and management together.

Antarctic Exceptionalism: Other Worlds, Cold Worlds, Cold War

Antarctic exploration diminished during the two world wars, although industrial whaling continued to expand in the sub-Antarctic during that period.[5] When extensive travel south picked up again, many of the states involved were caught up in the geopolitical maneuvers of the Cold War. The USSR and the United States devoted attention to developing their national presence in the Antarctic as a corollary to the technoscientific and nationalistic bravado of the "space race." The idea that Antarctica is in some ways the closest correlate to outer space on the planet helped scientist explorers convince Cold War governments to support their Antarctic projects.[6]

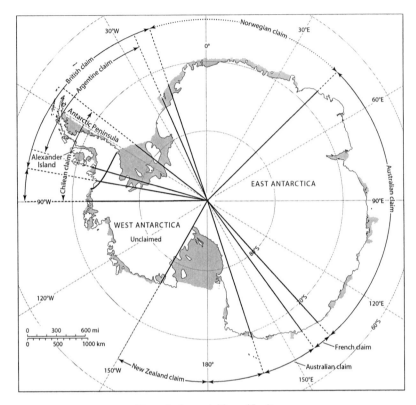

Figure 4. Antarctic Treaty Territory.

The other, more convincing, reason that the United States and the USSR gave for staking out outposts in Antarctica was the need for strategic positioning. They reasoned that perhaps a military presence in the far south would be advantageous in the unforeseeable and dangerous future. The possibility that pockets of mineral and oil resources existed in the polar south may have also motivated these nations to establish nationalistic Antarctic outposts. The desire to demonstrate technoscientific superiority was likely another impetus. To these ends, the United States built the Amundsen-Scott South Pole Station at the South Pole, Palmer Station on the Antarctic Peninsula, and McMurdo Station in the Ross Sea region, which continues to host the largest human population on the continent (1,100 people in the summer). Because of McMurdo's large size, significant infrastructure, and multiple runways, any location in Antarctica is accessible from this station. The U.S. presence at the South Pole gave the appearance that the nation had a stake in all previous territorial claims, except for Norway's. The USSR took an alternate approach: it built small stations within the boundaries of each other nation's territorial claims and a base at Lake Vostok, near the fabulously named "Southern Pole of Inaccessibility"—the most inland location on the continent. The three American bases are still in operation, but contemporary Russia has scaled back the number of its bases in favor of focusing attention on five key bases: Bellingshausen, Mirny, Novolazarevskaya, Vostok, and Progress.[7]

The nations that had claimed territory in the Antarctic—France, the United Kingdom, New Zealand, Australia, Argentina, Chile, and Norway— became increasingly concerned about the posturing of the USSR and the United States. The International Geophysical Year (1957–1958) helped focus attention on modern scientific endeavors in Antarctica and elsewhere, including space.[8] This event and the well-publicized scientific projects that took place under its mantle raised public attention about Antarctica in many nations. In this climate of public support for Antarctic scientific research, representatives of governments that were involved in or interested in Antarctica met to discuss potential Antarctic futures. The matter of national territorial claims and the buildup of Soviet and American infrastructure throughout the continent was the primary policy concern. The solution to this problem was the Antarctic Treaty, which was signed in 1959 in Washington, DC, and entered into force in 1961. Many of Antarctica's geopolitical and environmental dramas are based on interpretations of this treaty.

In the context of a legacy of deadly and glorious exploration and a rich amount of science fiction and "alternative science" imaginations about the possibilities for people and nations in Antarctica, the Antarctic Treaty formalizes the continent as a site that constitutes an exception to the forms of internationalism that have been created in other parts of the globe. The treaty codifies the concept that while Antarctica is a continent full of nations, nationals, and nationalism, officially it is nationless, or tenuously international.

Antarctic Exceptionalism: Indigeneity

Ethnographic research offers opportunities to learn more about some of the key concerns about human cultural processes. In my fieldwork, these opportunities are often centered on the notion that the Antarctic is exceptional as far as earthly continents go. A base employee at Scott Base pointed out the most striking Antarctic exceptionalism to me over beers at the government-subsidized bar. As we sat on our bar stools in the most densely populated area in Antarctica, a group of employees was wrapping themselves in sheets as they prepared to drive over to McMurdo Station for a toga party. He asked what my research was about, and I briefly explained my project to him. He looked at me and said: "But there are no people here."

Antarctica is the only substantial landmass without an indigenous human population. People deal with this Antarctic exception in some creative ways. The Scott Base employee considers the continent to be unpeopled, though our very presence there—which was made possible by the New Zealand government—suggests otherwise. The Argentinean government's approach to Antarctic indigeneity and nationalism has led to a unique Antarctic program. In all other national Antarctic programs, children are not allowed at the bases. If women living and working in Antarctica become pregnant, they are sent back to their home nation as soon as is feasible because they represent a medical liability and a potential source of public consternation and worry about the health of the mother and child. However, at Argentina's Esperanza base on the Antarctic Peninsula, entire families are encouraged to move south if a family member works there. The Argentinian government provides a school for these children.[9] In 1978, Emilio Palma was the first person born in Antarctica, at Esperanza base; he is sometimes labeled the "first Antarctican." That this indigenous Antarctican is also Argentinean is one of Argentina's rationales for its territorial claim.

Some scientists and environmentalists—particularly but not limited to people with long histories of time and work on and about the ice—make claims for themselves as indigenous Antarctic people. At a meeting I was a part of during the Antarctic Treaty Consultative Meeting in Edinburgh in 2006, one environmental activist spoke about how her job in the activist projects she pursues in other parts of the world is to publicize indigenous perspectives of the area, suggesting that these perspectives were powerful and authentic environmentalisms that needed to be given a public platform. Antarctica is different, she said, because there are no indigenous people to speak for the environment.[10] She suggested instead that Antarctic environmentalists must take the place of indigenous spokespeople and in some ways become indigenous Antarcticans.

This seemingly close relationship of environmentalism and indigeneity is reproduced elsewhere, and the notion of Antarctic environmentalism in general burgeoned with the growing popularity of the environmentalist movement in the West in the 1980s. One of the most critical and most frequently referred to Antarctic Treaty documents is the Madrid Protocol, which was written in 1991 and entered into force in 1998. This document attempted to standardize environmental management for all of the international members of the Antarctic Treaty. (I outline its specific history in the next chapter.) Essentially, the Madrid Protocol adds a third category—environmentalism—to the Antarctic Treaty's original mission of promoting "peace and science." In Antarctica, environmental management and specific kinds of environmentalisms rooted in Antarctic experience are seen as integral to the peaceful conduct of science. The Madrid Protocol and the ways signatory nations enact and interpret it in international meetings make it possible for specific kinds of policy to emerge and for these policies to be implemented in the Antarctic in particular ways. One of the key technical moves of the Madrid Protocol is that it includes the expertise of scientists and their research while simultaneously providing exceptions from environmental regulation for scientific research. The notion of science in the service of environmentalism is entangled with environmental policies in ways that both support and undermine these policies.

Antarctic Exceptionalism: Science

Antarctica is often characterized as environmentally, geopolitically, and culturally exceptional. The people I worked with during my fieldwork reinforced these ideas as often as they dismissed them. Science has a privileged role in Antarctica and in the Antarctic Treaty System. My colleagues almost always regarded one important component of Antarctic human life as an arena where exceptions could be made: the practice of scientific research. On the one hand, there is a unique situation in Antarctica in which self-regulation and the cultural norms of Antarctic scientific practice are what dictate the protection of the Antarctic environment. Antarctic scientists advise policy makers about environmental regulations in their areas of expertise, and the knowledge gained by Antarctic researchers, supported by their national programs, provide the political rationale and the policy suggestions that govern the continent. On the other hand, Antarctic scientific research projects can almost always obtain exemption from the regulations put in place to protect the environment. Nonetheless, scientists are bound by other ethical considerations than the ones listed in the Madrid Protocol document and many of these researchers exceed environmental and other standards mapped out for them, improvise practices in the field that reduce their impact on the environment or the species and things they study, and heavily pressure their colleagues to monitor and improve their impacts on the Antarctic environment and the residents therein.

Since scientists can usually find ways to argue that their research merits exceptions from the policies laid out in the Madrid Protocol, the most tangible Antarctic environmental policies are living practices, such as camping rules. While other Antarctic places, such as international meeting rooms, museums, and conference rooms, are seemingly interchangeable, the Antarctic environment matters, literally and tangibly. Scientists, base employees, and field support staff are trained before they enter Antarctica to organize their living situations and practices in relationship to their safety and to the strict environmental standards laid out by the Madrid Protocol. There are no exemptions to environmental regulations for camp and station living outside of research and emergency scenarios, so these everyday practices of living in the Antarctic become the focal point for enacting environmental protection.[11]

Some Antarctic science becomes the stuff of present-day myths and legend, especially so-called Big Science projects. For example, an international team of physicists is conducting Project Ice Cube at the South Pole; they are constructing the world's largest scientific structure, an under-ice neutrino detector that occupies one cubic kilometer. The United States Antarctic Program built an ice road from the coast at McMurdo Station to the South Pole in order to transport fuel and supplies for this construction project and other supply needs at the base. The building of this road has been tremendously controversial, particularly because "wilderness" in the United States is usually defined through its lack of roads. Building a 900-mile-long road to the center of the Antarctic continent raises concerns about the wilderness status of the region and about tourism and other increased human uses of a previously difficult-to-reach spot. Russian scientists have "penetrated" Lake Vostok, a subglacial lake with water that had never been accessed by people and may contain unknown biota. The International Polar Year (2007–2009)—a continuation of the International Geophysical Year fifty years earlier—inspired dozens of international scientific ventures with a heavy emphasis on Big Science, including large-scale ship exploration to map the marine biodiversity of the Southern Ocean. Charismatic research projects capitalize on popular conceptualizations of Antarctic science as work that is conducted in a remote, outer-space-like region.

Antarctic scientific exceptionalism—both the tremendous creative potential for obtaining new knowledge and the exceptions that permit scientists to both participate in and opt out of the environmental policies for the continent—is born from the same environmental imaginaries and historic legacies of exploration, extremism, and nationalism that frame the current geopolitical situation for the continent.

Extremes and Exceptions

Much of this chapter highlights the moments when the Antarctic continent and its people seem to stand apart from the rest of the world, even though life and work in Antarctica are always in conversation with the world elsewhere. At the same time, people who are intimately familiar with living and working in and about Antarctica also note the boring, rote, procedural, and bureaucratic ways that Antarctic life can be styled: this also borrows from a history of the involvement of the militaries and other governmental entities in creating and maintaining Antarctic research operations.[12]

Exploration histories and environmental imaginaries are anchoring points for the case studies in the chapters that follow, as are the particular formations of science and governance that occur in Antarctica. The imaginaries of the place and the ways human activities in Antarctica are treated as exceptional and mundane help create space for looking at the ways

Antarctic environmental policy is made at the level of the document and the implementation of documents in the field. Antarctica is imagined as haunting, colorful, and exceptional, and these visions become routinized as imaginations filter through drafts of international policy and are practiced as governance.

2

THE ENVIRONMENTAL HISTORY
OF THE ANTARCTIC

I spent the sunny December nights in Antarctica in a Macpac tent. The tent's rainfly was not quite up to the freezing and thawing occurring on the icy crust of the Windless Bight, off the landmass of Ross Island. Even though it shredded slightly as the wind bore down on it, my tentmate Holly and I styled comfortable temporary lives as Antarctic field researchers.

Holly and I each had our own vestibule near the shredding rainfly. We each dug the snow out of the floor of our vestibule about half a meter down and used the space for storing all sorts of things needed to live in Antarctica: two huge pairs of boots, jagged and hazardous crampons, duffel bags stuffed with extra layers of merino underwear, the ubiquitous pee bottle and fascinating female urinary director, and several different-colored trash and recycling bags in varying stages of fullness.

Most of the things we stored in our vestibules had been issued to us by Antarctica New Zealand employees in Christchurch as we prepared for our trip south. These items are a material record of a legacy of previous Antarctic expeditions, both successful and unsuccessful. Some equipment, such as the pillowed mukluk boots and the bulky overmittens, are copies of clothing worn on early Antarctic expeditions led by Scott, Shackleton, and Mawson, except that some are made of modern fabrics. Some of the materials are issued with concern for our comfort and safety while in the field, such as the clothing layering system and our crampons. Over time the people who use these items have tweaked them to include new fabrics and technologies.

Other items consist of environmental materials formally codified into the Antarctic field experience for scientists, support staff, and base employees. These materials relate to management and care of the environment—for example, the system of colored trash bags for sorting recycling, food waste, and other waste—implement certain requirements of the Protocol on Environmental Protection to the Antarctic Treaty (1998), informally referred to as the Madrid Protocol. This document attempts to articulate the Antarctic Treaty System's decisions about how to protect the Antarctic environment. It contains six annexes that more completely outline agreements on matters such as waste disposal and legal liability and it established the Committee for Environmental Protection, a crucial part of the work of the Antarctic Treaty System. The Antarctic Treaty System's view of the materials used in the Antarctic to manage the environment suggests that these items and their associated practices will optimize Antarctic environmental protection and will minimize human impacts on the Antarctic landscape. This chapter explores the histories of environmental practices in Antarctica described by scientists who conducted research there both before and after the Madrid Protocol. Environmental practices have changed with the implementation of the protocol, and most people see these changes as a definitive shift. The Madrid Protocol reconfigured scientists' technical and material relationships with the Antarctic environment.

I will also describe the ways Antarctic scientists with field experience before and after the enactment of the Madrid Protocol document frame their recollections of environmental practices over time and the ways they were complicit with the policies of the time or enacted their own environmental practices in ways that were not always in accordance with official policies, as written.[1] These more individualized practices are usually framed in terms of improving official recommendations; that is, those who follow them argue that these practices are more protective of the Antarctic environment than the protocols in place. From another perspective, some scientists' unofficial environmental practices engage with official policy by attempting to follow top-down practices but fail to do so because of their sheer absurdity in Antarctic field situations.

Environmental impact is mitigated by the practices of scientists but also, importantly, by the ways people and things are ordered and trained for survival in the Antarctic.[2] These environmental practices are the subject of this chapter. In particular, I will discuss the scientific practices and practices of everyday Antarctic life in field camps that are designed to minimize human impacts on landscapes, the sea, and wildlife.

The stories in this chapter rely on particular engagements with the forms that clear scientists to visit and conduct research in the McMurdo Dry Valleys. The forms are important both materially and as forms of idealistic discourse imbued with the best environmental and bureaucratic intentions. Anti-mining scientists, activists, and policy makers have created an idealistic world of Antarctic environmental protection. This world was temporarily made real during Greenpeace's "occupation" of Antarctica in the 1980s at its World Park Base. However, the world these forms seek to make does not entirely emerge when the scientists who have filled them in arrive in the Antarctic. This world on paper (policies and regulations) becomes "edited," and the imagined, idealistic world fails to materialize, or at the very least materializes differently than the policies intended. The scientists and policy makers in these stories agree that regulation through policies, laws, and conventions is the best way to effectively protect the environment and that these regulations are important and valuable. This chapter examines how important rules are made and the documents that were designed to bring about compliance and the editing that occurs when the rules simply do not work in the ways their makers intended.

Memories of Environmental Behavior and Policy Changes

Antarctic scientists (guided in various degrees by Antarctic decision makers) have shaped their living practices to minimize their impact on the environment since the implementation of the Madrid Protocol in 1998. Scientists who worked in the Antarctic before and after the implementation of the protocol almost always describe a story of increasingly strict environmental procedures for field camps. Several remember instances of what they would now consider to be environmental irresponsibility and tell stories of such instances when they recall how things have changed in the past few decades. These long-term memories of environmental impact illustrate the dramatic changes that occurred around the time the Madrid Protocol came into force. Practices that were environmentally suspect but were not regulated became unimaginable and illegal during the 1990s. A major question that many of my project participants raised in interviews centered on the authorship of Antarctic regulations: Were changes in practices the result of scientists' increasing concern about their research site and the Antarctic environment in general or were these practices determined from the top down as national Antarctic programs began implementing the regulations laid out in the Madrid Protocol? Project participants invariably touched upon the shift in environmental practices, offering several perspectives about who—scientists, managers, or policy makers—had coordinated this shift.

People working in the Antarctic prior to the Madrid Protocol recall many examples of environmental negligence and abuses of various sorts, including the routine tossing of garbage into cracks in the sea ice, leaving heavy machinery on ice to either break out to sea or melt through into the water, walking through rare moss communities, and disturbing Adélie penguin colonies. Others recalled detrimental practices such as bulldozing hillsides to create flat spaces on which to build McMurdo Station; allowing small oil spills to occur; losing gallons of fuel in the blank landscape; leaving gear and garbage to blow across the ice sheet; burying human waste only to rediscover it, uncovered, the next season; and various other practices. These practices were usually described in two ways. The first way relied on relativism; that was the way things were done back then and people did not consider it to be an issue. The second way described these practices as problems that the interviewee noticed and tried to mitigate by bringing the issues up with his or her national Antarctic program, especially when mitigation required serious logistics, such as flying empty fuel canisters out of deep-field camps.

Noticing and remembering these practices, though, indicates another kind of engagement with these former environmental practices that are now seen as egregious. The massive open garbage dumps, bases polluted with toxins, and unhealthy floral and faunal populations are part of an acknowledged and accepted history of the Antarctic. They echo Western environmental practices of the 1950s and 1960s elsewhere, just as the mitigation process laid out in the Madrid Protocol and carried out in the national Antarctic programs from the 1990s to the present echo environmental protection and management schemes in other parts of the world. The Antarctic environment is often considered to be more pristine than most of the rest of the planet, so the policies outlining Antarctic protection require strict adherence and attention to detail from everybody visiting the continent. For the most part there is international agreement about the protocol, but there are still some spectacular disagreements over what is environmentally acceptable in the Antarctic. These disagreements often intersect with other human activities in Antarctica related to science and logistics. For example, the U.S. ice road to the South Pole, the Russian plans to drill into Lake Vostok, and the opposition to making the Ross Sea a marine protected area indicate that in post–Madrid Protocol Antarctica, not everyone agrees about environmental policies. However, environmental scrutiny at all scales has become accepted and expected.[3]

Modern environmental management in the Antarctic stems from an international conflict over the Convention on the Regulation of Antarctic Mineral Resources (CRAMRA). CRAMRA emerged from concerns about resource extraction and the Antarctic Treaty's silences on that issue. The CRAMRA document outlined the Antarctic Treaty System's approach to regulating mining in the Antarctic, particularly but not limited to oil and natural gas exploration and exploitation. It allowed exploration for and exploitation of such minerals using the precautionary principle, or anticipatory management.[4] This document also explicitly gives international consent to activity related to Antarctic minerals, albeit in a strictly managed and strictly organized way. CRAMRA was negotiated from 1982 to 1988 and was adopted by consensus at the Antarctic Treaty Consultative Meeting in 1988. When a policy is adopted by the Antarctic Treaty consultative parties, the approved policy must be ratified by each of the signatory states, who then enact the international policy as domestic law.

As CRAMRA entered national policy agendas, environmental groups and Antarctic scientists, with the help of some sympathetic members of the press, lobbied to stop its ratification. Many people consider Antarctica a "last wilderness" or a scientific laboratory and were opposed to any activity related to minerals on the continent and its waters. In New Zealand, several scientists and policy makers I worked with during my fieldwork were involved in efforts to stop the New Zealand government from ratifying CRAMRA.

Sam is one such person. At the time he was a biology doctoral student; he is now a professor at a New Zealand university. Sam considers himself a "Greenie," which, to him, is exemplified through his refusal to own a car; he chooses instead to cycle everywhere about town. Following New Zealand deliberations in Parliament over CRAMRA in the 1980s, Sam felt that only a total mining ban would be sufficient in the Antarctic. To encourage the New Zealand Members of Parliament to refuse to ratify CRAMRA, he instigated a petition and letter-writing campaign (see text below). Sam sent letters to every Antarctic scientist in New Zealand, asking them to reply by signing the petition and writing their own letters. He also wrote key New Zealand policy makers and government employees involved in CRAMRA activities.

After I had interviewed him in his university office, he e-mailed me and offered to let me read through the files he had kept from his petition campaign. As I sifted through the stack of faintly yellowed papers, I recognized the names of several now-senior Antarctic scientists and policy makers based in New Zealand. Their letters ranged from outright support for a rejection of CRAMRA in favor of a total mining ban to those who supported CRAMRA for various reasons.

Text of Petition

I have undertaken scientific research in Antarctica and believe that:

1) commercial minerals activity is unacceptable because;

   —of the high risk of serious accidents and the inevitable environmental damage arising from the construction and use of onshore/offshore structures
   —the value of scientific research which depends on the unique qualities of Antarctica as a region largely free from the effects of human disturbance would be degraded.

2) great efforts should be made to replace the convention on the regulation of Antarctic Mineral Resource activities by a new convention which excludes commercial minerals activities and maintains Antarctica in its present near pristine condition as a conservation region in which only low impact research and carefully controlled tourism are allowed.

The letters people sent back to Sam, for the most part, supported his efforts. Several responding scientists mobilized both their scientific expertise and their privileged status as persons who had been to Antarctica. One signed the petition and wrote, "I would add that in my opinion as a geologist the search for minerals ashore and offshore will be fruitless in an economic sense."[5] Another scientist, a chemist, signed the petition and wrote,

> our findings over the past two years have indicated that the marine benthic communities in Antarctica are very sensitive to disturbance. Stable sediment regimes are crucial to the communities' existence. Any increase in rates of sedimentation or change in the grain-size structure of marine sediments would be disastrous. Such changes can come about easily from mining activities and would have immediate effects on the marine benthic ecosystem.

As experts with long-term experience in the disputed environment, these signers added scientific (and economic) observations to their opinions. While opinions are always shaped by the experience and status of those who offer them, the opinions of experts—especially those grounded in Western science—often convince non-experts.

One scientist who signed the petition typed out further suggestions:

> I have signed, *but* . . . since we are indeed scientists, I think we could be more constructive. For instance:
>
> —only a near-pristine environment can be used as a reference for monitoring of global pollutants, solar radiation changes, climatic trends and plant response to them, etc.
>
> —human disturbance is not measured only in terms of physical and chemical parameters: exotic microbes could have major effects upon slow-growing and delicate Antarctic ecosystems.
>
> —Global fossil fuel resources and easily-extractable mineral deposits must all sooner or later be exhausted. When that happens, alternative fuels must be used and minerals extracted from seawater or asteroids. It makes much more sense to invest the money needed to develop mining technology for the Antarctic in research on alternative fuels, synthesis of hydrocarbons, recycling technology. Or are our resource economists and politicians determined to apply false premises yet again to do the Antarctic what they have already done to most of our forests?

Underneath this typed-out text, which presages present-day concerns about the environment tin the Antarctic, he wrote in pen: "A petition like this will only have impact with politicians if it offers viable alternatives to development, i.e. if it assumes that there *is* a sustainable argument, based upon conventional economics, for Antarctic mining, and the petition then aims to contest the wisdom of that argument, and to put forward views which the supporters of minerals activity are attempting to over-ride." So in addition to his broad science-based environmental concerns about the Antarctic, this scientist also was concerned about the strategies for anti-CRAMRA engagement with pro-CRAMRA governments. Instead of simple opposition, he suggested that the anti-CRAMRA contingent— "since we are indeed scientists"—should offer alternative suggestions that would resonate with the people who were supporting CRAMRA.

A few who responded refused to sign the petition. Two who did so suggested that the preservation of Antarctica for the pursuit of scientific research would not convince many people other than scientists. One asked, "Who are we to try to preserve Antarctica as a scientific playground?" Another scientist who continues to speak out against management of the environment in the Antarctic to the present day did not sign the petition, adding that "in this case I will not sign it since the response will almost certainly be used by fringe organizations to support their unscientific and often emotive arguments." He paired his reluctance to associate, even anonymously, with "fringe" environmental groups with claims that such groups do not have scientific expertise and instead rely on more ephemeral tactics.

One environmental organization involved in anti-CRAMRA activism— though I would hesitate to call it a "fringe" group—was the international direct-action group Greenpeace. Greenpeace continues to be politically involved in Antarctic issues, particularly recently in its work to stop Japanese "scientific" whaling. (Japanese boats have begun to travel into the Antarctic Treaty waters on whaling trips.)[6] In the 1980s, Greenpeace began conducting ship-based Antarctic expeditions because of its concern that there were no independent observers of environmental conditions on the continent; only Antarctic Treaty states occasionally, and not systematically, observed other Antarctic Treaty states' environmental practices. These activist expeditions revealed images and reports of the egregious practices the scientists I interviewed mentioned.

Greenpeace relies on descriptive, often incendiary press releases and events designed to attract sympathetic media attention. When reports filed from the Antarctic expeditions did just that, Greenpeace parlayed this attention into another Antarctic endeavor. From 1987 to 1991, Greenpeace established World Park Base, located near the McMurdo Station of the United States and Scott Base of New Zealand. Greenpeace staffed the World Park Base with rotating teams of four residents. The name the organization chose for this year-round base represented what its members saw as the best alternative to the Antarctic Treaty System. Greenpeace promoted a world park run by the United Nations to manage the continent instead of the Antarctic Treaty System because membership and authority in the treaty system was achieved only through a scientific presence in the Antarctic—an expensive proposition for any state. Because McMurdo Station was the largest base on the continent (summer populations exceeded 1,000 scientists and support staff), it received most of Greenpeace's attention.

Greenpeace's four-year occupation of World Park Base generated many legends that environmental activists and their supporters love to recount. Newspapers and magazines ran stories about the Greenpeace base and activists' accounts of conditions and practices at McMurdo Station, the organization's direct actions against Japanese whaling ships in the area, and its trips to the French base of Dumont d'Urville to stage protests against the construction of an airstrip there. My interviews with scientists who were working in the Antarctic during this time also contain recollections of Greenpeace activities. For example, a New Zealander scientist recalls a day when Greenpeace activists staged a march through the "streets" of McMurdo Station. A few other scientists remembered that they or people they knew made side trips to World Park Base to provide the Greenpeace activists with accounts of environmentally problematic activities, often reporting on the activities of their national Antarctic program. In 1992, when the Madrid Protocol was approved at the Antarctic Treaty Consultative Meeting, effectively banning minerals activity, Greenpeace ended its continuous Antarctic presence and completely dismantled and removed World Park Base. While the organization did not help create a world park in Antarctica, its members saw the Madrid Protocol as an effective compromise that could enact the Antarctic environmental world they imagined and, for a time, had created through their actions on the outskirts of McMurdo Station.

Concerns about the Antarctic environment are in the minds of others besides cosmopolitan, Western environmentalists. The "Question of

Antarctica" is one of those bauble-like phrases I take out and think about every so often, a polar koan—because although much work has been done to answer the Question of Antarctica, it is a question that has not been asked. Instead, it is an agenda item at the General Assembly of the United Nations. Prime Minister Mahathir Mohamad of Malaysia brought "the question of Antarctica" to the UN General Assembly in 1983. He contended that although the Antarctic was a global commons, the Antarctic Treaty System treated it as a domain for research conducted by wealthy and privileged states. Mohamad used the Question of Antarctica to raise concerns about undemocratic governance in Antarctica and to consolidate his leadership within the nonaligned world (Hamzah 2010). Antarctic and Southern Ocean Coalition (ASOC) founder and environmental lawyer James Barnes helped Malaysia draft this text; governance concerns merged their interests. The question of Antarctica pokes at issues of the perceived exclusivity of the Antarctic Treaty and what should be required of those who conduct some of the most expensive scientific research on Earth if they are to have a say in the management of a continent. Malaysia became a signatory to the Antarctic Treaty in 2011, bringing the symbolic outsiders inside. The United Nations Environment Programme usually reports to the UN General Assembly on this question every three years, although whether the United Nations will formally pursue the issues the question raises is uncertain (Beck 2006).

In its occupation of World Park Base, Greenpeace did not always—or even usually—follow the increasingly strict and regulated environmental policies that emerged with the Madrid Protocol in the 1990s. Greenpeace does not always figure in versions of Antarctic history, as the story of the continent is now one of international cooperation in the name of peace and science. As a critic of nation-based Antarctic governance, Greenpeace makes some Antarcticans uneasy. Greenpeace's involvement in strengthening environmental regulations in Antarctica is often overlooked in favor of stories of international cooperation and scientific responsibility and innovation. Certain alliances are erased or considered to be less legitimate because they are seen as political rather than scientific or international. I do not discount the efforts of Greenpeace in Antarctica and suggest that its work helped forge an alliance in the 1980s between environmental activists, the press, and certain Antarctic policy makers and scientists. However, this alliance was not formal, or even explicit, among these groups. It is similar to the whispers I heard from scientists in my interviews with them about this time period: it was piecemeal and fragmented, unorganized, but bound by shared concerns (see Tsing 2005). Perhaps these allies did not even consider each other as allies: they did not share many of the same interests. This environmental alliance allowed states to take credit for the creation of the Madrid Protocol without having to acknowledge that other actors were involved in the creation of an Antarctic world with better environmental protection.

This sort of whispery alliance between some state representatives in the Antarctic Treaty System and environmentalists persists in the post–Madrid Protocol policy world. The meetings are peopled by state department or foreign service diplomats and staff, scientific experts, and lawyers. These parties have a vote (or, more accurately, a veto) in the decisions made in the meetings. Several environmental organizations are invited to the Antarctic Treaty consultative meetings each year as "expert" delegations who are able to supply opinions and information but do not have a vote in policy decisions. One group, the Antarctic and Southern Ocean Coalition, is the only nongovernmental environmental group that has access to the Antarctic Treaty consultative meetings.[7] ASOC is an umbrella organization that represents environmental NGOs concerned with Antarctic problems, such as the World Wildlife Federation, Friends of the Earth, the Whale and Dolphin Conservation Society, the Wilderness Society, the Sierra Club, and Greenpeace.

ASOC delegates work tirelessly in the meetings, in hallways, and in private meetings and during coffee breaks to foster alliances and agreements between their organization and state delegations.[8] Some national delegations have strong environmental positions, such as those from New Zealand, Norway, and the Netherlands. Australia, New Zealand, the United Kingdom, and the United States each invite ASOC to nominate an environmental NGO advisor to sit with their delegation, which gives the advisor more access to policy conversations.[9] However, many more delegations have mixed opinions and official position statements. It is not uncommon for an individual national delegate to collaborate, or at least sympathize, with ASOC's environment-focused positions while their national delegation as a body decides not to support ASOC positions. ASOC deliberately maintains these partial, unofficial collaborations with individual state delegates. Like the letter-writing campaign and petition in New Zealand, and the visitors to Greenpeace's World Park Base in the 1980s, Antarctic environmental activism relies on tenuous, informal, and quiet relationships between government officials, scientific experts, and environmental activists. ASOC is formally part of the Antarctic Treaty System in this understood but not explicitly acknowledged capacity. This provides opportunities for direct campaigning about environmental issues and oversight of the Treaty System and a way for the interpretations of an environmentalist NGO of Antarctic policy making to be made public through ASOC reports and press releases. However, the approach of letting environmentalists "in" to the meetings also silences some other, possibly more radical and hard-line environmental activists: ASOC is interpolated as less radical in the slow, relationship-centered policy-making process that dominates the consensus work carried out in the Antarctic Treaty System. This interpolation is an outcome of the Madrid Protocol.

The World on Paper

One scientist turned policy maker, named William, was a senior consultant for ASOC at the Antarctic Treaty consultative meetings and the Commission for the Conservation of Antarctic Marine Living Resources in the 1990s and 2000s. William is British by birth, earned his PhD in New Zealand, and now lives and works in Australia. His various home places are not atypical of an Antarctic scientist or policy maker; to be Antarctican is to be worldly, cosmopolitan, and often indebted to the travel networks of the British Empire. Intent and serious in relation to the environmental atrocities and lazy bureaucracy he notes in the Antarctic Treaty System, he is also jovial, often lightening the mood of meetings and workshops he attends and contributes heavily to. William is also well read in the history and philosophy of science, and his contributions to our conversations integrated broader histories of world environmentalisms, geopolitics, and cultural practices of international treaties.

In the 1980s, he was involved with anti-CRAMRA activism and recalls that during this time, "Antarctica was the biggest environmental issue in the world for a year." He and Sam were both graduate students at the time, and they collaborated on anti-CRAMRA events in New Zealand. Following this heady activism and the rejection of CRAMRA by several national governments, the Madrid Protocol began to be written.

The Madrid Protocol can be seen as much as a reaction to Greenpeace activism as it was a reaction to the proposed mining activities. While the protocol stopped mining activities in the Antarctic, it also preserved the integrity and stability of the Antarctic Treaty System as the governing regime for the continent. William says that the Madrid Protocol was drafted during an expansive period of what he describes as "second wave environmentalism."[10] However, he feels that the interpretation and implementation of the Madrid Protocol has been carried out in a more conservative political climate.

The document is not all original text. With a touch of irony, William claims that the writers of the Madrid Protocol "mined CRAMRA for good ideas" for their document. The environmental impact assessment process, which I describe in greater detail below, is taken in large part from CRAMRA text. The Waste Disposal Annex, which enacts the system of waste sorting I described above, was initially a recommendation of the Scientific Committee on Antarctic Research (SCAR) to the Antarctic Treaty consultative meetings. In fact, each annex of the Madrid Protocol, except for Annex IV on Marine Pollution, was developed elsewhere and inserted into the new document.

One major change the Madrid Protocol brought about in the Antarctic Treaty System was the creation of the Committee for Environmental Protection (CEP). The CEP meets for most of the first week of the Antarctic Treaty consultative meetings. It deals with problems and policies relating to all of the annexes of the Madrid Protocol and is one of the liveliest meetings within the ATCM. It is not without problems, however. While most Antarctic Treaty states have signed the Madrid Protocol and are therefore CEP members, participation during the meetings is dominated by a few states, most of which are Anglos (in Antarctic Treaty parlance), or English-speaking state delegations. Also, the CEP has become increasingly bogged down by its massive agenda, as all development activities (such as the construction of new stations) and management measures (such as proposals for Antarctic specially managed areas) must be deliberated on in the space of four or five days. While some annexes, such as Annex V on Managed and Protected Areas, are just beginning to be interpreted in the CEP, others, particularly practice-oriented annexes such as those on waste management (Annex III) and environmental impact assessment (Annex I), are becoming engrained in the everyday life of national Antarctic programs.

The scientists I interviewed described Madrid Protocol regulations as increasingly strict rules relating to camping practices, waste disposal, recycling, and interactions between humans and wildlife. Some of these practices, especially among the New Zealand scientists I spoke with, had been adopted as "common sense" or "good manners" decades before they became international policy. Some national Antarctic programs, including that of New Zealand, implemented practices the Madrid Protocol mandates before 1998, but now all bases and field camps are expected to use specific means to minimize impact. For example, most researchers I interviewed talk about institutionalized environmental practices, such as recycling, the handling of gray water waste, and toileting practices. Essentially, all human-origin waste is to be sorted in the field and treated and pumped into the sea (for some liquid wastes) or packed up and returned to the nation responsible for it (the latter is what happens in most cases). Field camps are to be set up to minimize the human footprint to the "greatest extent practicable," which varies across national Antarctic programs. These strict, policy-driven environmental practices designed to minimize the impact of human waste in the Antarctic follow a Western "leave no trace" ethos familiar to many outdoorspeople.[11] However, because not every Antarctic scientist and field support or base employee is fluent in the micro-managerial principles of leaving no trace, such practices have been written to suit tented field camps in the Antarctic and engrained with varying success into the organizational cultures of national Antarctic programs and the international Antarctic community.

Another significant change the Madrid Protocol brought about involves carefully configured interactions between people and Antarctic wildlife. People are instructed in pre-departure videos and handbooks to leave a distance of ten meters between themselves and any wildlife they encounter. (This is informally called the "ten-meter rule" in the New Zealand Antarctic Program.) This distance is designed to prevent stress among seals, penguins, and other Antarctic fauna. There are exceptions to this rule. Scientists doing research involving animals may be permitted to approach the animals to take samples or to "harvest" some animals as specimens for their research. Another exception is that when animals approach humans, the ten-meter rule is not enforced. (Sometimes people try to mimic animal behaviors in the hope of getting an animal to approach them.)

Not everyone involved in the changes in environmental practices since the Madrid Protocol describe them as orchestrated by environmental activists and international policy makers. Many do, noting that preliminary environmental evaluations (PEEs) became more scrutinized and stricter year by year, suggesting more specific practices for the details of living in and conducting research in Antarctica. One scientist I interviewed named Stuart took an opposite tack, suggesting a grassroots shift among Antarctic scientists with regard to their environmental practices. Stuart is also one of the few scientists I spoke with who credited Greenpeace efforts with bringing about changes in environmental practices in the Antarctic. As we sat in his Christchurch office, I asked him what his observations were about how environmental practices had changed over time and who was responsible for these changes. Stuart told anecdotes spanning some decades to support his theory that specific environmental practices, including the decision to stop hunting seals to feed sled dogs (since dogs were no longer allowed in Antarctica) and the wholesale cessation of jumping into the Dry Valleys lakes, a customary rite of passage for New Zealanders in that area. He said,

> There have been some spectacular changes. The Antarctic is only a small sideline of my other work here . . . but I use the Antarctic as an example. . . . It's almost like a grassroots movement of environmental change that wasn't really driven by bureaucracy. And there were practices that used to take place, as I recall, in the Antarctic that people just one day said we're not doing that anymore, and these kind of changes weren't driven by any directives, and it was really interesting watching this. And I think probably the single most important thing that caused these changes was Greenpeace. Because suddenly it made people aware of things that they'd just overlooked before. And in the early 1980s, both Scott Base (but not quite so much as the U.S.) and the U.S. philosophies were almost like running these bases like military outposts. And even the New Zealand program, they used to choose ex-military people as their operations managers and so on, even though there were no military people per se there. But that's how they sort of thought that it should be run and those were the kind of people they choose. So that kind of environment was one of looking after people and logistics and resources first and environmental matters came very far second. While we could never leave toilet waste or anything around, but things like gray water for instance we used to dump in the field. I have memories and I'm sure there are photographs around if you want to dig around of people dragging heavy metal waste out in front of Scott Base out into the sea ice and it used to melt through. All garbage used to get rid of by incineration but heavy metal waste was towed out onto the sea ice and it melted through.

Stuart's perspective on the history of environmental management in the Antarctic is anchored in his personal experiences on the continent. His narration also highlights his specific positionality related to the Antarctic. It is one of a highly masculine, adventurous, individual who through his experiences comes to shape a new approach to living in the Antarctic in an independent and grassroots way. Although it was international policies and the resultant national legislations that stopped the activities he described above, his lived experience of this shift is dominated by a sense that scientists determined which environmentally detrimental practices to stop and did so without the intervention of national laws or international decision making. I do not know if this is because of a specific technique through which New Zealand Antarctic managers implemented their policies or if the policies were almost seamlessly in sync with scientific environmental practices and opinions at that time or if indeed scientists initiated these practices in the New Zealand Antarctic Program and then these practices later became translated into the policy circuit, where they were formalized for the rest of the national Antarctic programs. Stuart grounds his accounts of environmental policy making in the Antarctic in his understandings of Antarctic scientists as very environmentally concerned researchers, to the point that they would sacrifice components of their research in favor of strict environmental protection. Further, he suggests that the policy makers and regulatory bodies (in his story, the National Science Foundation) had to curb the protectionist zeal of particular scientists so scientific pursuits in the McMurdo Dry Valleys could continue:

> I was involved with two NSF workshops, invited over by NSF to participate in these, and I think this is the classic one. That one is called the Environmental Management of a Cold Desert Ecosystem and the other one that you can get from the NSF is called McMurdo Dry Valleys Lakes: Impacts of Research Activities. And in that three-year period [between the papers] we developed a sort of a code of conduct, really, for operating and that eventually became the code of conduct for the ASMA, the Specially Managed Area for the Dry Valleys. My impressions of this one, and you can check with [editor] because he edited this, is what we were looking for here was we were looking at every single research activity that could impact someway adversely on the lakes of the Dry Valleys. And there was amazing detail about what should and should not be done. And there were no bureaucrats at this meeting except for some people from NSF but there was no one from

the State Department or anything like that there. My impression of that meeting was that the scientists got really carried away with environmental impacts themselves. They put so many constraints on their own activities by their concerns for the environment that I think—and you need to check this with [editor]—but the National Science Foundation actually asked us to tone down the language in here because it was so restrictive on scientific research activities. There were some parts of research that you might not have been able to do. And it was the scientists themselves doing it with no prompting. And that's what I was trying to get at as this sort of philosophical change in people's thinking about what they were doing in Antarctica. I'm on the editorial board of a journal . . . and I remember writing an article once as an editorial on that thing which really, it said something along the lines that now with this environmental imperative that's put on research you really need to think about whether you should be doing any research down there at all now. So what this has done I think is to put some real constraints on scientists to say "Is the research that I'm getting out of Antarctica more valuable than the environmental damage that I'm going to be doing there?" And that kind of thinking doesn't apply in any other area I know.

While most scientists I interviewed did not agree with Stuart's perspective that scientists created and followed their own environmental regulations in the absence of or counter to state control, neither did most describe the shift as an entirely top-down, policy-driven change. Instead, most scientists speak about the change in environmental practices of the Madrid Protocol time period as an engagement between themselves as experts and the policy makers. This engagement was not always smooth and easy but was characterized by creative arrangements among people, documents, and the Antarctic environment.

Arranging Rocks

> Where rocks are moved for campsites or other activities, they should be replaced in their footprint if possible, and at minimum should be placed with the salt-encrusted side faced-down.
>
> ENVIRONMENTAL CODE OF CONDUCT FOR THE McMURDO DRY VALLEYS

Near New Zealand's Scott Base and the McMurdo Station of the United States, the McMurdo Dry Valleys region is windswept, dramatic, and free

Figure 5. Tented field camp near Canada Glacier, McMurdo Dry Valleys. Courtesy of
Peter West/National Science Foundation.

of snow. This landscape is a polar desert that attracts photographers, scientists studying earthly equivalents of Mars, and environmental managers hoping to preserve this extreme and unique place. The appeal of ice-free areas such as the McMurdo Dry Valleys means that a relatively high number of people want to visit, and human visitation impacts the environment. The McMurdo Dry Valleys and other ice-free Antarctic places are defined by their surfaces. In the ground, particularly among rocks, nematodes, lichen, and bacteria persist in this seemingly lifeless landscape. These species are some of the most best-adapted creatures to extreme and otherworldly environments. The potential human impacts are many and difficult to quantify; they include fuel spills, noise pollution from helicopters, contamination of ancient lakes, and the effects of human activity on the ground itself (such as compaction, erosion, or the making of trails).

When the Madrid Protocol was implemented, certain environmental practices became regulated by national Antarctic programs. I will use one of these practices as a case study—the very mundane detail that a field party must resituate any rock they move in ice-free areas to approximately the place it came from—to discuss how Antarctic scientists engage with practice-driven policies. The official rule that one must resituate rocks is located in the Environmental Code of Conduct for the McMurdo Dry Valleys, a set of field rules given to all visitors to the McMurdo Dry Valleys Antarctic Specially Managed Area. This international set of guidelines that outlines behavior between rocks and scientists is not about a trivial matter, though it can seem so compared to dramatic Antarctic environmental events such as large blocks of ice plunging to the sea as the continent melts. The practice of restoring rocks to their original location is related to those much more dramatic images. Antarctic scientists who measure the flow of disintegrating glaciers in order to explain the planet's environmental problems try to fashion their lives in Antarctica in response to these larger stories they tell. When they restore rocks to their original locations, they are taking seriously the Antarctic environmental problems they are the spokespeople for.

When a field party arrives in the Dry Valleys, the members set up their camp in a cluster of tents for living, sleeping, and working. In an ice-free area, people move rocks for two practical reasons: to clear a space for tents and to use them as equipment. Antarctica, especially the Dry Valleys region, is extremely windy, so scientists use rocks to weigh down the lightweight, temporary materials associated with camping. Antarctic rocks can anchor down tent flies, the lids of toilet buckets, and various other pieces of equipment that can take flight when the winds pick up.

Most of the scientists I interviewed agree with the practice of putting rocks back where they came from. However, many who followed this practice often singled it out to highlight some tension between what is on paper in the policy and what takes place in the Antarctic field.

The practice of restoring Antarctic rocks to their original locations literally begins on paper, when principal investigators fill out a suite of forms prior to their group's departure south. One document is the preliminary environmental evaluation, or PEE. National Antarctic program environmental officers use this form to try to achieve uniform environmental practices at a high standard, either by complying with the Madrid Protocol or by following a stricter policy. The PEE form is where a scientific team officially documents what its practice will be with regard to moving rocks.

Susan, a glaciologist, said that her experiences with rearranging Antarctic rocks began with her involvement with filling out her group's paperwork and her willingness to follow the practices she described in the documents when she was in the field. However, she found that the political realities of Antarctic research made possible practices that were different from her group's practices and ethos. Susan said:

> So when I first started work in the Antarctic, you know, I had to fill in forms and explain what I was going to do and all that sort of stuff. I remember the first time I went down I filled in my preliminary environmental evaluation form and it came back and it said what are you going to do with the stones after you take your tent down? And clearly the expected answer was, I am going to roll them back to where they came from, so I said I would roll them back to where they came from, and indeed that's what I planned to do and that's what I did. I didn't mind doing that, I didn't mind doing it at all, because I agree with minimizing the environmental impact. The . . . thing that confounded me down there was that, there we were rolling the stones onto the side of the tent and then rolling them back, putting them back to where we got them from and that was all fine. It was slightly disconcerting to find that there were great big cans that had been set up many years before, before any of these practices came in, whether they're cultural or environmental is another matter. But then what was disconcerting was the Americans came back in one year and laid some concrete for a weir and it was like, hang on a minute, I'm rolling my stones back. . . . It wasn't concrete, it was sandbags, but substantial, near-permanent structures. When I got back to New Zealand after a couple years it was, like, hang on, Antarctica New Zealand isn't going to get very far with expecting people to roll stones about if the Americans are going to come in and put down concrete and set up these structures to look like they're there for twenty years. You can't reasonably expect people to sensibly engage with that difference. I remember thinking about this because there aren't many situations on the globe where you have people operating in the same area under different sets of rules because of national boundaries. You go to France and you have to do what French law says whether you're from America or from New Zealand. And I remember discussing with the environmental manager at the time . . . talking to her about this issue. And discussing how silly, from the point of view of someone on the ground, it was just a bit disheartening when you're doing your best. And I think to a certain extent that was one of the stimuli for the Antarctic Specially Managed Area plan that's now in place for the Dry Valleys. Trying to get all the people who are operate there, whatever country they're from, operating from the same set of rules with the same stringencies.

Susan feels that the solution to a problem in field practice is to be found in reiterations of policies on paper that try to steer scientists toward uniform environmental practices. In this narrative, Susan is a proactive editor of practical environmental policy; her engagement with the practices of her national Antarctic program centers on her suggestions to managers. Her story also contains an anti-colonial narrative, an oft-repeated, joking complaint about the extravagances of the U.S. Antarctic program in comparison with the more humbly funded New Zealand Antarctic program. Antarctica New Zealand is fairly reliant on the United States Antarctic Program to provide most of its transportation south, both the flights that carry scientists and staff to and from the field and the annual resupply trip that relies on icebreakers. New Zealand, in turn, provides a critical logistical base for the United States Antarctic Program in Christchurch. This portrayal of national Antarctic program operates something like Latour's "centers of calculation," in which an empire's center engages in "re-creating distant parts on paper and in collections . . . [requiring] that things distant be rendered 'mobile,' 'stable,' and 'combinable'" in physical sites such as Kew Gardens (Miller and Reill 1996, 23; Latour 1987). The center in this case is in the practices and offices of environmental managers who sift through field reports and management plans, matching environments and local political situations and adapting and translating these plans for other places and people. In short, the way Antarctic environmental practices are conducted, through intensely material, fragmented, and technical processes, renders Antarctic policy piecemeal. Antarctic policy is much more effective at producing minute technical standards than it is at producing sweeping mission statements for the entire continent, though policy makers do both.

Gretchen, a PhD student, talked about the difficulties of following this rule with what she considers to be full compliance. She contends that following this policy to the letter would take a lot of time and use technological apparatuses, so she has reached a compromise in her handling of the rocks at her field site: "We don't put our rocks back exactly where we found them because we have a hundred rocks that we would have to remember the GPS coordinates of . . . which is a bit ridiculous, but we do put them back in the general area that they came from." Many scientists who talked about moving Antarctic rocks suggested that this involves simply setting the rocks back from where they came in a general sense and that the rule existed primarily to preserve the aesthetic and wilderness qualities of a relatively highly visited, easy-to-impact Antarctic area. Gretchen, who has read the Code of Conduct, considered the geologic impacts of removing the rock from the dirt matrix of the Dry Valleys floor, noting that her moved rocks did not make it back to the precise geographical coordinates from where they came. Furthermore, she considered the orientation of the rocks to the valley floor, placing the stones back to where the rock interacted with the soil, forming layers of "salt." To Gretchen, how a specific rock surface was associated with the surface of the valley was more environmentally important than the precise geographic origin of the rock. While she seems to make a concerted effort to replace rocks she has moved, she does not comply with the rule. The complicated dictates and impossible practicalities of a rule about a seemingly small detail of Antarctic camping seems to lead to reinterpretations and foster violations.

Another scientist, a British-born geologist named Laurie, grappled with the logistical problems that the policy of rearranging rocks brings to the material reality of Antarctic tent camping. In her interview, she wondered about the practicality of this rule and engaged briefly with an evaluation of the environmental impacts of rearranging rocks. Laurie contrasted the value of mitigating impact on the landscape—restoring rocks to their original location in order to preserve the image of the Dry Valleys as an unpeopled environment—with the ecological impacts of moving rocks. There is a gap between recreating the image of what the area looked like before people camped there and the effects on the Dry Valleys surface at the molecular and microscopic level. Merely re-creating the image can be considered a bad faith effort, but this is usually also considered good enough. She says:

> The thing that was quite a bit different was the emphasis that they put on the amount of rearranging of rocks that you could do around your campsite to secure your tents and things like that. Well, you have to be very minimal, so you have to go back to preexisting campsites if there are any in the area that you want to go. And if you have to remove rocks to secure things, then you have to put them back exactly where you find them. And you have to be very careful walking around your camp that you don't disturb the ground too much. We didn't do that at all when I first went down. It's very impractical. I can sort of see why they suggest it but it's very, very impractical when you're using rocks to anchor your tent or anchor your toilet buckets and that sort of thing or to just to put on your gear to stop it blowing away. You know it's been done like that traditionally because you don't want to load up

your helicopter with heavy things to stop things blowing away when there's rocks which you can easily move and use once you get there. But to do it properly to have minimal disturbance you've really got to take heavy things with you to weigh down your stuff when you get there, it's really just not practical. And you can take your rocks and put them back to where you got them from, but the ground's still being disturbed. I'm not quite sure what they're trying to achieve, if it's visual impact they're trying to reduce or if they're truly concerned with ecosystems or whatever it is, but if that's the case then they're being disturbed by moving it at all.

Laurie describes the potential for rocks to be used as equipment for camping in the Antarctic field. Their presence in the landscape prevents valuable fuel and helicopter time from being used to transport other materials to use as anchors.

Laurie reflects on her confusion over the reasons why this policy exists, particularly when using the rocks is so practical. Laurie and some others suggest that the fact that a detail such as restoring rocks is included in national environmental policies and the agreed-to rules for Antarctic Specially Managed Areas is often indicative of a failure of bureaucrats in Antarctic programs to understand what Antarctic fieldwork is actually like. While many Antarctic policy makers in national programs and in nongovernmental organizations have a science background, usually a PhD in a scientific discipline and fieldwork experience in the Antarctic, Laurie's statement highlights a division between scientists and bureaucrats.

In addition to engaging with sensibilities about pragmatic economies in field camping and situating herself as a "field" person in contrast to desk-bound bureaucrats who write policies that rule out sensible field practices, Laurie mobilizes her scientific expertise regarding the ecological problems of resituating rocks.

Another interview participant, a geologist named Howard, is concerned about the effects of moving and replacing rocks on the landscape and on the geology of the Dry Valleys. He says:

In 1957, when we first went into the Dry Valleys, it was a pristine place. It had |only| been flown over during a previous Byrd expedition. They photographed it . . . so absolutely everything was virgin. So the landscape of rocks was thousands to tens of thousands to hundreds of thousands of years old, the surface was very old, it's pretty unique in any part of the world. It's a great pity we did not have the rules then. . . . The rocks have been moved around so much. It's not just moving the rocks but walking on the rocks, so in my view the damage is done. I just wonder if you can find really pristine areas there that you can be absolutely sure no one has walked across. . . . If you have any experience in landscapes and you go into the Dry Valleys and you walk around or you fly around, the whole place is moved around. The Dry Valleys . . . are deserts, so they are a pavement of ventifacts and rocks that are armoring the surface and all you have to do is pick one rock up and what's underneath it but powdery silty sand, in fact. So when you go in and put a tent in, you clear all these rocks away, you put your tent down to get that nice soft silty sand, which is better to put your sleeping bags on. So all the rocks get put in a rock circle: no one puts those rocks back. And if people do try putting rocks back they're not putting them back in the right place.

Howard uses his scientific expertise in geology to depict the impossibility of successfully resituating rocks in a way that prevents impact on the desert pavement. He also deploys two scales of long-term history: the thousands of years of geologic history of the McMurdo Dry Valleys landscape and his personal history of experience in the Dry Valleys, one of the longest personal engagements with the place on a continent with only about 200 years of direct human history. In his experience, many people simply do not restore rocks in Dry Valleys camping areas, and when they do, the practice fails. Howard's two long-term historical perspectives leads him a somewhat pessimistic assessment that no place in the Dry Valleys is untouched by people and that the practice of putting rocks back to their original locations does nothing to preserve the geologic integrity of the landscape's surface.

These four interviewees spoke about one small detail of an extensive suite of environmental practices codified in international policy. There is play within the rock-resituating rule that enables people to talk about it differently and to interpret it differently in the field. These four scientists chose to make rock resituating about prevention, about mitigation of impact, or about minimizing damage. In each of the interviews I did with them, they described how the intended world written down in the McMurdo Dry Valleys environmental policies fails to materialize and is edited to appear only partially or not at all.

These four people, along with many other Antarctic scientists, deal with both the material and discursive realities of practice-based policies.

They engage with the paperwork and associated bureaucratic scrutiny of the preliminary environmental evaluation process and then enact these policies in the Antarctic environments where they conduct their field research. They inscribe policy both onto paper and onto the environments where they work. Their decisions to fully comply or to comply but find a policy ridiculous or to not comply and assume that no one else does either illustrate their understandings of the relationship between policy and practice in an extreme and protected environment. A pragmatic decision about what to do with rocks demonstrates both the merging and the disjuncture of expert and lay perspectives (Henke and Gieryn 2007). These interview participants edit the policy of restoring rocks through sophisticated understandings of the limits of bureaucratic enforcement, the realities of Antarctic field work, and expert knowledge of Antarctic environments and the impacts people have on them.

Finally, the substance of the rock as an artifact also matters. Rocks are not inert symbols from which human dramas are enacted and negotiated. Instead, as John Law advocates, rocks and other nonhuman forms "should be seen as forming an integral part of such systems. . . . Their form is thus a function of the way in which they absorb within themselves aspects of the seemingly non-technological environments" (Callon et al. 1986, 237). The regulation about handling rocks in the Antarctic fosters multiple interpretations and transgressions. The highly technical practices required for "perfect" rock replacement in the McMurdo Dry Valleys are nearly impossible, and this impossible regulation enables field scientists to engage in discourses and material practices that shape the Antarctic landscapes they work in.

Making Worlds on Paper

The history of Antarctic environmental practices, from the egregious base practices of the 1950s and 1960s to the furor over the Convention on the Regulation of Antarctic Mineral Resource Activities to the adoption and then interpretation of the Madrid Protocol, has brought specific Antarctic environmental worlds into being. First, concerned organizations and individuals in a specific, pro-environment political era imagined that Antarctica was a place in need of specific environmental protections and regulations. Next, from this idealistic period came the Madrid Protocol, which itself is a formal and idealistic act of world-making that explicitly spells out in fine detail a strict level of environmental practices and mechanisms. This highly regulatory world implements the precautionary principle and involves a high degree of international oversight. The environmental activist "fringe" groups (meaning, loosely, people who are not officials or Antarctic scientists) have been included in this world by being invited to the Antarctic Treaty consultative meetings (and the Madrid Protocol's Committee on Environmental Protection), but they are somewhat marginalized. In addition, the world created in the Madrid Protocol is not completely brought into existence as it is interpreted and practiced in the Antarctic field. As I illustrate above, scientists in the field substantially edit the regulation requiring that Antarctic rocks be restored to the spot from which they were removed. In this editing, the people who are instructed to resituate rocks invoke scientific expertise, histories of irreversible human impacts on the McMurdo Dry Valleys, and pragmatic views of the use of time, labor, and equipment. Thus, the Madrid Protocol regulatory environmental world has not yet fully appeared. However, the protocol is the best device for successfully protecting the Antarctic environment at this time. The policies detailed in the Madrid Protocol make it possible for field workers to "save" the Antarctic at the same time they are degrading that environment, since they are both the primary enactors of strict environmental practices in the Antarctic and the primary source of human-caused degradation there.

One of the most salient lessons from the example of the regulation about restoring Antarctic rocks has to do with how interviewees' understanding of history influences their contemporary rock-handling practices. Interviewees configure both history and practice in the quotes above as either determined by or responding to policy. Scientists and policy makers engage with the Antarctic environment in productive relationships that creatively grapple with the related environmental policies.

The work of writing policy and the negotiations, meetings, and re-writings that comprise "writing policy" center on implementation and interpretation as much as on writing: policy editing takes place as the policy is enacted. Policy makers and scientists interpret each other's work throughout this writing process. Shore and Wright (1997) describe policy as an "intrinsically rational, technical, action-oriented instrument" and claim that policy contains models of society (5–7). They argue that while these models are often bound up with moralistic discourses such as environmentalism or cultural conservatism, moralistic policy makes such decision-making "more generalized, more impersonal, bureaucratic, and anonymous" (10–11).

Policy that is perceived to have these attributes is usually considered successful because the more political aspects of a piece of policy writing have been neutralized. Simultaneously, however, such policy also consolidates power (Ferguson 1990). One way of making Antarctic policy seem neutral is to downplay overarching rationales and to focus on technical and material practices that turn moralistic rationales into new embodied relationships with the Antarctic environment. As I have illustrated, the scientists who implement these top-down environmental policies interpret them as they practice them in the field. As experts, these scientists also describe the ways the environment may respond to such practices, explaining the geologic or biological effects that moving rocks, sort of restoring rocks to their original location, or not doing so at all may have. There is thus some tension between Antarctic scientists and policy makers about who can plausibly be a spokesperson for environmental management and protection, with the most credibility given to those with strong technical sets of knowledge and histories of practical experience in Antarctic environments. Both technical knowledge and practical experience are enfolded into the discursive practices that result in Antarctic policy writing.

The practices outlined in the Madrid Protocol have been taken up by Antarctic scientists, either very directly through training and the scrutiny of environmental evaluations or indirectly through the organizational culture of national Antarctic programs and the traditions of Antarctic research. The scientists I have worked with have noticed a shift in practices in the past decade or so that involves stringent technical practices designed to minimize and mitigate environmental impacts. The Madrid Protocol and the shift in practices in the Antarctic field reflect changes over time in response to histories of environmental ethics elsewhere, people's changing relationships with the places where they work, and a policy climate that encourages productive inscriptions of environmental practices, scrutiny, and editing.

The way scientists edit policies is shaped by their disciplinary training. These lenses determine how they relate to Antarctic rocks; the environmental meanings of the rocks are made clear through the disciplinary sets of knowledge of the field scientists. Scientists use their training to evaluate and change their practices in relationship to these subjects and to the policies managers hand them.

Regulations are the best technique available for protecting the environment. In Antarctica, idealism exists in policy making as the sense that in Antarctica, of all earthly places, environmental management can be different, stricter than in the rest of the world. In some instances, while the world the Madrid Protocol created emerges—trash is shipped off of the continent, preliminary environmental evaluations are filed—it is a partial emergence. This world fails to materialize in some micropractices, such as rock resituating. There is enough world making to acknowledge the rock policy and enact it in paperwork, but the policy-mandated practice is edited in the field.

3

SENSING THE ICE

When I camped on the Ross Ice Shelf, I donned my mukluks every evening when we finished our observations and travels for the day. The mukluks for Antarctica New Zealand are blue, pillowy boots reminiscent of
moon boots. They are not fit for skiing or walking since they lack support,
but they are soft and warm, an Antarctic equivalent of slippers to wear
around your campsite.

Our group was alarmed, however, when the ice shelf underneath us started
melting. It was midsummer and temperatures hovered around freezing,
which made our extreme cold weather gear much too hot for most of everyday
Antarctic camp life. A slight tick above 0 degrees C rendered our entire landscape soggy and heavy and water seeped into our tents and clothes. As the snow
melted, our mukluks seemed to melt as well. As we padded around the trails
we made between our tents, the wet mukluks left blue imprints in the snow.

Some of my companions suggested scooping up contaminated snow and
taking it back to the base. I had done this the first morning, when I had tripped
over a guy line and spilled my coffee. But as the steps proliferated, that task
seemed impractical. However, more than seven years after this Antarctic trip,

I remember the blue footprints, the concern over leaving residue in the snow, and the seasonal warming that made the footprints visible and possible.

These tracks had nothing to do with the environmental monitoring research we conducted while in Antarctica, but they were part of the phenomenon of field research in which we experienced everyday life while conducting the formal rituals of scientific practice. The tracks flagged one of the central concerns about the future of Antarctica: in a warming world, how will the massive Antarctic ice sheets behave? How much and how quickly can the melting ice sheets change global sea level rise? While the footprints in the slushy top layer of snow indicated seasonal warming rather than a sustained global warming, they nonetheless inspired us to think about the insubstantial nature of the substance we were living on.

In this chapter, I analyze the other ways scientists come to know about the ice as they go about their work, sensing it both formally and informally. To do this, I read my ethnographic field and interview notes for the moments when scientists depict Antarctic ice casually and intimately, as people who know the ice in ways that most people do not. I will describe field research in the Antarctic and the technoscientific encounters and misadventures my interview participants experienced. Antarctic terrain is sometimes difficult to work on: scientists sample ice that melts, try to drill into unaccessed subglacial lakes, or must negotiate moving across a surface that shifts and melts. I will also look at how data from the field becomes ice-sheet models and at some of the sensory and experiential impressions of human interactions with such models. My analysis will reflect on expert elicitation panels, a highly structured, subjective means of characterizing expert opinions on uncertain topics. Finally, this chapter will discuss how these subjective and personal experiences help shape experts and the judgments they make about the future of the ice sheet. Relationships on the borders of discipline and expertise are mediated through sensory experiences with data, the alignment of Antarctic research with nations, and the epistemic scale-making that grounds all scientific practice.

Sense, Experience, and Knowledge

Although scientific ways of being in the world are diverse and contingent on particular subjectivities, including disciplines, they are disciplined in particular ways. Maintaining the ideal of sensing the world in an objective, precise, replicable, and truthful way is a complex task. While techno-scientific apparatuses make remote sensing possible, offering a perspective beyond that of simple human capabilities, scientific sensing takes place at all scales, including the scale of intimacy (Helmreich 2009). Much of the philosophy that helps us understand how scientists achieve this is filtered through the lens of epistemology (classic ethnographic examples include Gusterson 1996; Knorr Cetina 1999; Rabinow 1996; Traweek 1988). Studies of how knowledge is produced look at the intrinsic and extrinsic ways people work with the world, and with data, to make some coherent sense of it.

That individual scientists bring particular subjectivities to their research is well understood. Indeed, particular subjectivities can be considered talent, an ability to see nature in a novel way (see, for example, Grandin and Johnson 2004; Haraway 1989; Ingold 2000). Evelyn Fox Keller titled her biography of Barbara McClintock *A Feeling for the Organism* (1983), referring to McClintock's often-repeated advice for how to elicit the truth from nature, even in the face of contradictory scientific knowledge. For Keller, McClintock's insistence on intimately knowing the plants she worked with—as individuals, not as specimen types—is a "longing to embrace the world in its very being, through reason and beyond" (199). Keller suggests that McClintock's groundbreaking scientific insight was inspired by an intimate knowledge of plants, knowledge so intimate that it was sublime and subliminal. The particular materiality of McClintock's research object was revealed through her repetitive, individualistic, and consuming expertise.

Scientific knowledge can be formed through intimate interaction with the objects of study, a methodology that contrasts with the ideal of a scientist as a detached, objective observer. For example, Hustak and Myers (2012) argue that in Charles Darwin's experiments with orchid reproduction, he "participated actively with his experimental subjects, to such an extent that he *moved with and was moved by* them" (85, emphasis in the original). Darwin's experimental practices involved sensory engagement with his objects of study. His findings were obtained only through encounter and relationship with the nature he was studying.

Scientific understandings of nature rely on the global scale. Tsing (2005) argues that contemporary environmental politics introduce processes of generalization that merge the global scale with "universal Nature" (90).

Scientists tend to be trained in this idea: to be local is to be subjective or anecdotal. Climate science, which is pinned to the global climate model, uses the global scale to imagine the future. "The global scale takes precedence," Tsing writes, "because it is the scale of the model" (103). The technological tool meant to make human thought about the future possible also shapes the scale at which thinking takes place.

Sensory experiences of the environment are easy to find in the ethnographic record. Many studies focus on indigenous people's relationships with the environment, revealing, for instance, the complexity of some indigenous understandings of ice across time and in local contexts (Carey 2010; Cruikshank 2005; Orlove 2005). These native knowledge systems are based on evidence and experience and imply a long-term intimacy with the known environment. Raffles (2002) analyzes how "local knowledge" is relational and intimate, knowledge that is laden with "affective sociality" (326). Rebecca Herzig (2005) developed her analysis of scientists' self-sacrifice, including self-experimentation, as a form of relating with the world, with truth, and with knowledge. Following this tradition, this chapter examines how scientists walk the boundary between sensory, intimate experience and objective, universal knowledge as embodied people in the world. My research also makes apparent how the embodied and the sensorial is not primitive, elemental, basic, or instinctual but is instead bound up in the complexities of nationalism, scientific translations of scale, and boundary skirmishes over what counts as expertise from within scientific disciplines.

Below, I analyze some of the relationships Antarctic glaciologists (including field scientists, modelers, and those working with remote satellite data) have formed with the Antarctic glacial environment. I do so in order to contribute to understandings about the tactile and experiential nature of scientific expertise, even when the experts claim to know little scientifically. Expert extrapolations about nature in the absence of data give light to field scientists' intimacy with the environment they work with and the virtual future worlds modelers interact with. This is why expertise and expert opinion matter. It is also why the physical, material, personal, and embodied characteristics of scientific expertise matter. My analysis considers the future of the West Antarctic Ice Sheet as the object of scientific concern that has inspired speculative, future-oriented research and the influence of sensory experience on expert perceptions about what might happen to it in the coming centuries.

Doing Science in Antarctica

In many ways, conducting Antarctic research is like doing the work of science anywhere, with some logistical adjustments for the cold, wind, and isolation. However, the relationship between the nation and science here is explicit and direct; nations and their contractors run research stations and permit access to the bases and logistical support. There are sixty-eight national research stations in the Antarctic run by thirty states.[1] Because of their origins and their current management, these stations often have a paramilitary atmosphere.

Stations are the logistical hub for researchers. Support staff live there and help prepare research teams for field trips. While some researchers stay at the station, most venture out into the field to collect data, staying in tents or small structures helicoptered to the research location. Some conduct deep-field expeditions that require flights or multiple days of land travel to reach and are more logistically complex. Often, researchers pack up samples and specimens and transport them back to the research team's labs in their home country, though there are exceptions to this. Some samples and specimens are analyzed in field labs, which are often constructed from shipping containers and helicoptered in, or in more permanent lab facilities at the research stations. Whether Antarctic materials travel back to "the world" or are analyzed in Antarctica depends on many factors, including time and weather sensitivity, whether the researchers have traveled south by ship for an entire research season or have been flown in for a shorter time, and the capabilities and equipment of the sponsoring national Antarctic program.[2]

Antarctic scientists conduct research in this remote place under extreme environmental conditions and alongside more familiar political, bureaucratic and material realities. While the mundane bureaucratic nature of much of Antarctic logistics does not often garner much attention from scientists, it nonetheless informs how they move about the place, conduct research, organize their camping practices, and survive in the Antarctic. To confront the sensational Antarctic environment and landscape is to engage simultaneously with wildness and civilization. The expert knowledge that emerges is a product of this engagement.

Crevasses

While the threat of ice melt and disintegration from climate change is often considered in terms of sea level rise elsewhere on Earth, field scientists relate to the risk of the shifting Antarctic terrain in a more intimate sense. Most of the human activity in Antarctica takes place on and around the ice sheet that covers most of the continent. The ice sheet can be up to a mile thick, but it is not solid. It moves, shifts, and cracks.[3] Every person living in Antarctica receives mandatory training on camping, first aid, and emergency survival skills so they will be able to cope with the cold, the wind, and the disorienting landscape. Crevasses are one of the hazards to be considered, although people working in the Antarctic tend to be modest about the risks they face and matter of fact about some of the dangerous experiences they have had. Bragging about one's near-fatal escapades is simply not done, at least among one's Antarctic peers.

Crevasses are deep cracks in the ice sheet. They are often masked by snow and are thus difficult to detect in the white, monotonous landscape. Moving across the ice requires keen observation and, sometimes, climbing harnesses. Near tented field camps and bases you will find flagged routes, areas of ice and snow that have already been surveyed as free of crevasses and simple to move across. But if you travel off a flagged route, you must be with a partner, be roped to that partner with a pair of climbing harnesses, and be ready to anchor that partner in case he or she falls into a crevasse. Motorized vehicles, such as the tractor trains that drive the route between McMurdo Station and the Amundsen Scott–South Pole Station, dynamite crevasses to create a safe road for transporting cargo. Landing aircraft on the ice sheet is particularly challenging, since some crevasses can be difficult to see, especially from the air.

In an interview, glaciologist Robert Long noted the difficulties of studying a location about which so very little is known, in rough terrain where one might unexpectedly have a closer-than-anticipated encounter with the ice sheet.[4] The long-term field study of the Siple Coast region of the West Antarctic Ice Sheet that he participated in involved a massive team of scientists and support staff who flew out to a remote field site each summer. Moving the drill and other major equipment to Antarctica involved over a year of planning, as the heavy pieces had to be shipped in containers from California.

Once the team and equipment arrived at McMurdo Station, they would load up a Lockheed C-130 (Hercules) plane, a large workhorse military vehicle capable of landing on ice. Pilots and researchers would fly over their proposed landing sites, taking aerial pictures to try to map out the crevasse patterns before landing. One time, however, the crew of the loaded-up C-130 failed to see a crevasse on Ice Stream D and landed in it. Everyone on board survived, but the plane was severely damaged when the engines hit the ice.

This event triggered a massive recovery operation. Approximately thirty mechanics worked on fixing the plane for two months, then winched it out of the crevasse and flew it out of the site. That season's drilling time was reduced from three months to one. The incident also changed the practice of landing C-130s on "unimproved" snow and ice, a decision that reconfigured how and with what equipment and field support glaciologists could work in the deep field. Now the much smaller DeHavilland Twin Otter aircraft supports deep-field projects, and large stationary projects have come into favor. Since the extremely bulky hot-water drills cannot be transported into deep-field locations any longer, glaciologists now must study the ice sheet with different techniques and equipment and in new locations closer to base. This safety- and risk- oriented decision limits locations to where research is practical instead of where it is most likely to yield significant data.

As in any field situation, the Antarctic environment makes particular research projects possible and impossible. Along with the complicated logistics operations that deep-field glaciologists have to contend with, the landscape they try to survey is literally shifting, from stable to deadly. Crevasses dramatically remind one that a landscape is not a thin line on top of the earth; a landscape is multidimensional. The multidimensionality of the Antarctic ice sheet, particularly the marine-grounded ice sheet in the west, is a tangible reminder that action in the environment takes place in surprise bursts and in unanticipated directions. For scientists trying to predict the future of the ice sheet with limited data and models, anticipating the ice's surprising multidimensionality is a key problem.

Melting Data

Though scientific knowledge is sometimes depicted as free-floating truth that a scientist manages to capture, scientists know that research objects are subject to environmental and human contingencies. Ice samples are precious and are obtained only with time, care, careful planning, and major logistical support from governments, militaries, and contractor services. Ice coring is a technical feat that involves drilling through hundreds of meters of ice in a way that cuts the sheet while keeping the ice intact. Once the ice is extracted it must be transported to a lab to be analyzed. These labs contain precise sawing machinery, carefully managed temperature and humidity conditions, visualization equipment, and machines that can analyze the small bubbles of air trapped in the ice. From the ice cores, glaciologists can read the history of the ice sheet, including its age and behavior over time. The air bubbles trapped in the ice contain the history of the earth's atmosphere, which scientists can analyze.

When Marc Cash conducted his PhD research in the 1980s, he made rudimentary observations at Ice Stream B, then took an ice core and prepared it so the National Science Foundation could ship it back to his lab. He wrapped the core in plastic, then in fur, and finally in a cardboard tube. That April, Cash got a phone call from his lab around 6 in the morning informing him that his shipment had arrived. Cash, in his characteristic colorful language, said "I go screaming down to the office and I haul the ice off the freezer truck into [his advisor's] cold room and . . . I open it up and there's these little refrozen puddles of ice with fur in it. Melted down and they just refroze it and shipped it." As Cash put it, "my PhD melted." Cash and his advisor were able to mobilize their networks of colleagues to obtain ice cores in the next Antarctic seasons with support from the National Science Foundation, but the momentary panic of losing an entire season of research made an impression.

While Latour (1999) took painstaking effort to describe the translations between lab and field, in Cash's case, the translations failed on a major scale, to the point that the nature as data became unintelligible. Here, melting ice symbolizes the challenges of reading nature from a distance, in laboratory settings instead of the field. The role of the nation, mediated through the National Science Foundation, explicitly winds through this tale as one of developing knowledge capacity for the nation through funding and complex logistics. While the data read from the ice may be permanent, the material itself can be destroyed before scientists can read it. Relying on the intactness of an almost intangible, ethereal, and temperature-dependent substance lends a materiality to ice-core research that relies on a careful managed laboratory (and transport) microenvironment.

Penetrating the Ice

In addition to coring the ice sheet, scientists interested in what lies beneath the ice can drill down and observe the subglacial water systems, soils, and topography that sit on the Antarctic continental bed. Several projects, both historical and contemporary, have aimed to penetrate the ice sheet so scientists can gather knowledge about the physics, hydrology, and biology of the environment under the ice. These projects help develop theories about the future stability of the ice sheet in a warming world. For example, the scientists of the Russian Antarctic Program penetrated subglacial Lake Vostok by meters and then centimeters until their drill reached the lake under the ice. Over the years leading up to and during the drilling,[5] delegates at the Antarctic Treaty consultative meetings expressed concern about the environmental repercussions of plunging into a lake using a process that put gallons of oil into a drill column. In response, Russia submitted detailed environmental assessment reports to the Antarctic Treaty consultative meetings over a period of years in order to gain consensus support for penetration. The discourse around this project involved both narratives of discovering a new earthly frontier and discussions of how to properly mitigate the risk of contaminating unknown lake biota.[6]

As governmental and public concern mounted over the oil-filled Lake Vostok drill hole, the technology of drilling and the ways scientists approached drilling the ice sheet changed in light of environmental scrutiny. Scientists now know that they are not drilling through pure ice onto rock but are potentially drilling into microbiospheres unique to the planet. In addition, the U.S. National Research Council has published a report on environmental stewardship of the Antarctic subgacial environment (2007). In the 1990s, according to one glaciologist I interviewed, "you could drop any kind of sensor, you could do tracer studies by pouring salt water or some dye tracer into bore holes, and nobody would blink. But it was actually discussion around exploring the large isolated subglacial lakes (or as they were thought [of] at the time as being isolated) that turned on the switch in terms of, well, if we want to study these environments, what about when we try to penetrate them?" The idea that one was sitting atop microbiological communities that were interconnected proved instrumental for rethinking how to approach scientific research in terms of environmental ethics and acceptable scientific methodology.

In practice, drilling is mundane and routine. In an ethnographic case study, Skyrdstrup (2012) analyzes the anticipation of a drilling team on one of the domes of the Greenland Ice Sheet. He fuses the workaday experience of field camp life with the global scale of producing knowledge for climate models through the drill. At Skyrdstrup's drilling camp, the cook makes fun of the scientists and their imprecision, everyone places bets on when they will reach bedrock, and the primary investigator pragmatically shifts her methodology to manage the risk of the borehole collapsing. The people in this camp know ice, and for them, it is neither neatly sacred nor profane.

Scientific knowledge about life under the ice sheet changed how scientists treat the ice and do their work on it. The discovery of new life forms and flowing systems of water in the subglacial environment helped scientists reconsider the ice sheet as a habitat and as something influenced from below. In short, interaction with the ice sheet reconstituted how humans regard and approach the icy monolith and the imaginative and scholarly possibilities inherent in it, including nationalistic posturing over knowledge frontiers and the discursive and material practices of discovery.

Modeling Data

Researchers use models that rely on various scenarios of future emissions to predict what will happen to the ice sheet in the coming centuries. Since I had focused much of my ethnographic work on Antarctic field scientists, I was surprised by the sensory appreciation ice modelers had for their models and for the physical ice. In an interview with one ice modeler, Trent Smith, I was struck by what motivated him to work in the Antarctic: he did not do so he could travel to and live in the Antarctic but because of the mathematical and physics challenge of making ice-sheet models. He summed it up by comparing modelers to glaciologists: "Most glaciologists

are climbers or outdoor people, and they like that, and then they get interested in what they're climbing over. . . . But I don't want to go for a walk
on it . . . that's not the motivation." Initially interested in ecology, he went
to university and "spent all my time chopping up rats," searching for something more compatible with his interests. He found it in an environmental
science course. He remembers one lecture in particular about energy transfer and solar radiation. "They were showing equations . . . and something
clicked," he recounted. "I could see that the equations reflected something
that you could observe and that was happening in reality, and that got me
very interested in the use of models and maths to understand the environment." When Smith started applying for PhD positions, he simply looked
for modeling—modeling of anything—in the title. His doctoral work involved modeling ice sheets, but he wanted a break from academia by the
time he completed his degree. Thus, he began his professional career as an
engineer who modeled urban sewer flow in a government agency before
eventually returning to ice-sheet modeling.

Modelers joke about their nerdy reputation compared to those of the
apparently more adventurous and athletic field scientists.[7] One said: "The
classic modelers, you know, I've got spectacles, for example. . . . A typical
modeler is introverted and not particularly practical . . . whereas the typical field worker is very much a can-do person and craves excitement. If
you're an outdoor person and like climbing mountains, you've obviously
got a slightly different outlook on what you want from life than a modeler,
someone who enjoys solving differential equations." He went on to describe the power dynamics between the two groups, since modelers often
consider the role of field scientists in the Antarctic to be to that of collecting
data for the models. Field scientists, in contrast, are often looking not for
basic data but for exceptional, dynamic events. He explained that "modelers want field people to analyze something that's pretty predictable so that
they can put it into their models, whereas field workers want to go and find
something new and exciting and essentially unpredictable." This difference in approach puts the acts of scientific analysis and scientific observation at odds with each other. The kind of looking a scientist wants to do is
related to disciplinary ideas about truth, discovery, and even what makes
an appropriate research narrative.

The delineation between field scientists and modelers is not always so
stark, however. Richard Cash, a glaciologist who works in both modeling

and field sciences, does not find the division to be as cut and dried as his peers do. His perception calls to mind the porous (albeit policed borderlands between field and lab that Kohler (2002) analyzes in his history of biology. According to Cash, "It's interesting, there are real field people who wouldn't crunch an equation and there are real modelers who don't do much in the field. But a whole lot of the people who are doing it are shaded somewhere in between. The . . . [names early glaciology programs] tradition really was you go out and make the best observations and then you turn them into the best model." Cash's career spans both field science and modeling, to the point that he could not remember which team he played on in a soccer game between the modelers and the "data" (though he does recall that he scored a goal). His use of the word "data" to stand in for field scientists, who gather data among other, more complex activities, is telling, though used in a joking manner.

How modelers interact with their models and data is well studied (Edwards 2010; Lahsen 2005; Oreskes et al. 1994). While this work analyzes how models come together, my research focuses on how models do not come together and how modelers live and work with their always-partial computational representations in pragmatic ways. Models of the ice sheet turn it into a series of equations, and modelers sometimes slip from talking about the behavior of the ice sheet into mathematical language. For example, while discussing the difference between ice sheets (which sit over land) and ice shelves (which float over water), one modeler explained that

> they flow in a very different way, these two ice masses. Ice sheets flow by—because they have traction at the base, so they in fact flow by what is called shearing, so they flow by layers of ice which shear on top of one another while an ice shelf is like you would drop oil on water. It spreads in all directions. It stretches and it spreads. It's floating on the seas. This is a totally different way of solving the force balance than solving for the flow. This leads to fundamentally different mathematical equations.

Here, the modeler turns the ice into an equation to be solved. His role is to provide calculations to fill in the gaps where observations haven't explained how glacial physics work. In other words, modelers create approximations that keep the model running.

Another individual who is both a field scientist and a modeler is uneasy about approximations in numerical modeling. He prefers working with what he calls simple models that are not run into the future but instead can be ground-truthed in real time. He told me, "Numerical modeling is always kind of making the best of it and it's what I don't like about it. So I'm more like a slightly old-fashioned scientist and want to do simple models. I'm more analytical with mathematics so you know exactly what you're doing, though you have your limitations." This researcher avoided the more speculative possibilities of future-oriented models and instead worked within the confines of simple models.

A remote sensing specialist also underscored the limitations of modeling, particularly ice modeling, which has a long way to go in terms of reflecting the glaciological processes that are occurring. As we sat in his office during an interview, he said that the "ice-sheet modeling community has been living in its sort of fake world for a long time . . . maybe modeling square ice sheets and square ice shelves because they didn't have much else to work with." That the models turned the ice into "squares" instead of the natural Antarctic topography illuminates the complexities of trying to create standard models from a haphazard, random, and contingent world. That the terrain cannot be depicted with realism raises the question of how the processes driving ice behavior are being simplified, computerized, and modeled. The picture of the future ice sheet that is being presented is limited, partial, or even inaccurate.

Modelers are practical about the mathematical, physical, and computational limits of their models as programs that represent nature, with the obvious caveats about limitations and unknown and inadequately understood processes. One modeler described thinking about nature and limited knowledge with this example:

Modeler: The trick is that you don't know anything about the water, really. You just know that it's wet at the bottom [of the ice sheet]. And then there's a number which is called the sliding viscosity, which relates the sliding velocity to the forces acting. And then this sliding velocity, *you've hidden that you don't know anything about the water*. So it's possible from satellite data to work out what the sliding viscosity is, or make a reasonable estimate of it. But you don't know how it's going to change as the amount of water changes, so you either pretend that it's not going to change, or you use some kind of untested model to say it's going to change by such and such an amount if something happens upstream and there's more water coming down underneath the ice stream.

Interviewer: So, like an educated guess.

Modeler: It probably would be nice if it was an educated guess. (My emphasis)

In the light of scrutiny that might interpret such statements as "proof" of a vast scientific conspiracy about climate change, it is important to think about the context the modeler is referring to. In this case, the modeler is "fudging," or making an approximation that connects chains of known calculations that are backed up with data drawn from observations. In complex systems modeling, the pieces of natural processes must fit together. In the absence of observations, something must stand in to make the entire model run. These "known unknowns," where some equation stands in for a knowledge gap, create uncertainties in the model, but these uncertainties can be described, quantified, and marked for future research. While the modeler I quote above candidly describes an approximation, it must be noted that scientists consider it preferable to know that one is fudging an equation and being open about that fact in one's methodology than unwittingly ignoring a process in the model altogether.

Nonetheless, the interim calculation that modelers choose to approximate a "known unknown" in their model is a guess based on immersion in the data of a modeled ice sheet, an understanding that is both expert and partial, and an attempt to work with their state of not knowing something along with what they know well (or at least as well as any other expert in their area). With approximations, modelers are working on the edges of what they know and do not know. These edges are negotiated through the modelers' numerical, data-rich relationship to the model and to their pragmatic understandings of the model's always partial reflection of the world it attempts to describe.

Expert Elicitation

During my ten years of ethnographic research with Antarctic and climate scientists, no scientific activity has been as internally controversial as expert elicitation, another technique for characterizing expert impressions of the unknown. Expert elicitation is a research method that systemically surveys experts to characterize well-informed opinions about a topic in light of missing data or high uncertainty. Subjective and personal, this practice gives some scientists an impression of soft data, of social science. Nonetheless, expert elicitation projects have done what models and semi-empirical methods have not yet been able to do: describe with some precision the likelihood of rapid disintegration of the West Antarctic Ice Sheet (see Pfeffer et al. 2008; Rahmstorf 2007). This concern about the future of the ice and the related global sea level rise has provoked several decades of glaciological research in the Antarctic. This future haunts daily practices of research and logistics in the Antarctic.

Since the 1950s, glaciologists have marked the West Antarctic Ice Sheet (WAIS) as a site of concern. This concern includes the geological instability of the ice sheet, since much of it sits below sea level and therefore interacts directly with the warming ocean. Scientists are also interested in the potential for meters of global sea level rise the ice sheet holds. In addition, the complex geophysics needed to explain ice-sheet behavior seem to stand just slightly outside of expert comprehension; there are still some mysteries to how the marine West Antarctic Ice Sheet works at a fundamental level. For example, modeling difficulties have made it challenging to project the likelihood of a rapid disintegration of the ice sheet, which would have major impacts on global sea level rise. Two recent publications (Joughin et al. 2014; Rignot et al. 2014) are the first to suggest that the ice sheet has already passed the point of no return that has put disintegration into irreversible motion. The future of the ice is difficult to know and major amounts of human effort and resources are poured into attempts to know it scientifically.

Expert elicitation about the WAIS came about with the collaboration of David Vaughan from the British Antarctic Survey and John Spouge of Det Norsca Veritas, an insurance company specializing in risk assessment. The findings of their project were published in *Climatic Change* (Vaughan and Spouge 2002), a well-regarded climate science journal. Vaughan and Spouge went through several iterations of trying to gather enough information to perform an adequate risk assessment. First, they looked at published literature, but there simply were not enough publications to be significant. Next, they tried to use statistical models—similar to the use of actuary charts for life insurance—but again, there just were not enough numbers to put together a credible statistical model. Finally, the researchers decided to use the Delphi approach, which has been described to me in an interview as a method "where you ask people whose judgment you think is better than average." Clearly, the decision about who to ask and what is considered judgment is remarkably fraught and subjective in the most basic sense of those words. The experts Vaughan and Spouge selected presented their sense of the likelihood of two scenarios. In the first scenario, the sea level will rise by twenty centimeters per century over the next 200 years. The experts felt that the likelihood of this scenario was high and that the likelihood of the second scenario, a five-centimeter sea level rise per century over 200 years, was low. Each expert also indicated how certain he or she felt that each scenario would occur. In the end, virtually no scientist thought the ice sheet would collapse in 100 years, but most people thought it could disappear in 1,000 years. But for the period of that 900-year spread, the researchers found little consensus (Vaughan and Spouge 2002).

Some people who work on the WAIS have reservations about expert elicitation projects, and there appears to be a division between those who see studies that use this methodology as a reasonable stopgap and those who find such research to be sloppy and feel that the findings cannot say anything credible. One modeler was heavily critical of expert elicitation projects, asking me to turn my recorder back on after an interview to speak at length about his opinion on the record. He called expert elicitation "a poor man's solution if you don't have anything else. If you don't have good scientific arguments, ask our friends what they think!" To him, the subjective methodology and human vagaries of expert elicitation constitutes a decidedly unscientific way of polling scientists about matters they have expertise about. It suggests that when scientists cannot use their conventional techniques to seek answers, "gut instinct" might have value.

The Intergovernmental Panel on Climate Change (IPCC) is a massive climate assessment that produces reports approximately every six years. It is jointly run by the United Nations and the World Meteorological Organization, and hundreds of scientists participate in writing and reviewing its reports, which focus on the physical science basis of climate change; the impacts of, adaptation to, and vulnerabilities of climate change; and climate mitigation. In the 2013 IPCC assessment report (called AR5), the authors decided to include expert elicitation publications for review despite the controversy about this method. The authors may have been trying to represent more of the available information than they had in the past; the 2007 attempt to represent the potential that the West Antarctic Ice Sheet would disintegrate rapidly was considered insufficient by both the scientific and the policy communities overall (Oppenheimer et al. 2007; O'Reilly, Oreskes, and Oppenheimer 2012; O'Reilly 2015).

For example, the expert elicitation research of Bamber and Aspinall (2013) received several mentions in the chapter on sea level change in the IPCC's 2013 report. The IPCC authors cite Bamber and Aspinall to underscore the probability that grounding line[8] retreat will contribute to global sea level rise through 2100, uncertainty over whether recent changes in the WAIS are the start of a long-term trend of disintegration or are simply regular variability (1185), and the "wide spread" of sea level rise projections based on expert elicitation, which indicates "a lack of consensus" on the probability of ice-sheet collapse (1186). In short, the IPCC used Bamber and Aspinall to highlight uncertainty and lack of knowledge. That the IPCC authors chose to include expert elicitation to a serious degree at all can be interpreted as a change in the types of information the physical science–based Working Group I of the IPCC is willing to consider. However, it used expert elicitation research in this case to underscore uncertainty and not knowing.

Bamber and Aspinall's article (2013) emphasized a quantifiable finding from their expert elicitation research. While noting that expert opinion is both "very uncertain and undivided," they wrote that "we find that the median estimate of such contributions is 29 cm—substantially larger than in the AR4—while the upper 95th percentile value is 84 cm, implying a conceivable risk of a sea-level rise of greater than a metre by 2100" (1). The IPCC's AR5, which was published after Bamber and Aspinall's article, provided model-based estimates of sea level rise that ranged from 28 to 98 centimeters by 2100. The mean ranged from 44 to 74 centimeters for various future climate scenarios. These numbers are from only the likely range (with a 66% confidence interval), not from the worst-case scenario. In other words, the model outcomes provide more dramatic projections of sea level rise than expert elicitation studies do.

Some scientists express concern that expert elicitation provides subjective results and may skew projections toward the more dramatic side of potential futures. But this case underscores what I and others have termed "erring on the side of least drama," analysis in which experts underpredict or downplay results (Brysse et al 2012; Hansen 2007). I suspect that this may have to do with long-standing values in the culture of science of being "modest," objective, and rational (Haraway 1997; Shapin and Schaffer 1989). The practice may also be a response to contemporary climate change contrarianism.

In addition, the research community studying the West Antarctic Ice Sheet is small, and findings such as those from various model runs might be informally circulated among experts before research is published. Conversations within peer networks, presentations at conferences and meetings, and opportunities for peer review allow some research findings to gain legs before they are published. These conversations and informal circulations of research are part of the subjective and intimate work of experts.

Though the methodology of expert elicitation is easily open to critique, particularly by the experts who participate (or who choose not to participate), the idea that scientific experts may make judgments that are "better than average" lets us think about how expertise is formed and about how the formation of the expertise, not just the resulting data and analysis, may contribute to helping us understand some of the unanswered or little understood questions about nature and its potential future trajectories.

Knowing Antarctic Futures

In his laboratory, Robert Long concluded our interview by talking about climate change. He recalled talking with Al Gore in the cafeteria of McMurdo Station, telling him how Antarctica seemed to be changing in the decades he had spent as a researcher there. When he first began traveling south, he had no idea about climate change, "then suddenly in 1996, I saw water running through McMurdo, and then later I saw the dripping water into the shear margins. Then I went to the Trans-Antarctic Mountains and there was water pouring down the slopes into these outlet glaciers from East Antarctica." Long suddenly had a concrete impression, like my group did with our bleeding mukluks, of the icy continent melting. The tactility of the melting, of living in it, dealing with equipment in wet places that are supposed to be frozen, makes an impression, even if the phenomenon is an isolated event not related to climate change. Like the news stories and extreme weather events in our home places, we cannot help but wonder if these things we have never seen before are part of a broader phenomenon.

In this chapter, I analyzed the ways glaciologists sense the ice, build predictions, and deploy anticipatory ways of explaining their scientific research. I discussed how scientists fill in gaps in their predictive capabilities with expert knowledge of the ice sheet that is both intimate and technical. Scientific research is planned around the hope of witnessing and studying natural phenomena; futures are imagined through conducting research and living in the Antarctic field. It must be noted that experts are doing the imagining in this case. How does expertise influence these visions of the future? In the examples above, expertise extends beyond epistemological matters when embodied, everyday experiences that accrue alongside the practice of scientific research are taken into account.

This chapter has focused on how scientists consider living in the world they work in. Their ways of being in the world envelop more than their disciplinary foci, and their narratives about these experiences can help us understand the totality of scientific experience. Sensing ice sheets is a complex activity. These massive and chaotic formations have challenged scientists to find a vantage point from which humans can attempt to understand them. For example, in one interview, a remote sensing specialist who also does primary field observations notes that the interior is a "binary landscape, snow and sky and that's all you see." But on the coast of the Greenland Ice Sheet, "You see the whole thing. You see calving glaciers. You see icebergs. You see a much more dynamic part of the ice sheet." Dynamic glacial events, such as collapse, underscore the potential of the ice sheet in spectacular fashion. Witnessing glacial activity at any scale involves situating oneself in relationship to the ice. There is a relational perspective to satellite observations as well; they provide a god's-eye perspective of the continent and its subglacial dramas. This observational work, both the work that is done professionally and the observations that occur as a consequence of living in the field, may serve as expertise or bias or as anecdote for scientists when they are asked about their sense of the ice sheet's future.

Observational data must be put into models to make the models as realistic as possible. As described above, models are simply complex computer programs that are coded to mimic the earth's systems. Since these systems are imperfectly understood, it is essential to underscore that climate models are simply approximations. At each juncture where a system is poorly understood—how the WAIS will melt, how the ocean will expand as it absorbs CO_2, or how the radiation from clouds protects the atmosphere or magnifies warming—the modelers write code that is an approximation of an approximation. These are mathematical fixes that allow the computer code to run, but they do so without any numerical representation of nature at all. However, modelers cannot provide the high-resolution projections needed if there is high uncertainty about the behavior of the ice sheet or a simple lack of data overall. Thus, even at a high resolution, the models contain approximations.

In response to several inquiries from governments and NGOs about the future of the WAIS, scientists have made projections. How did they do this if the model codes contain huge runs of approximations? They have formally and informally assessed the data at hand and related that to their experiences with the ice sheet. This is where the intimacy of field studies comes in, where people live on the ice sheet, gather pieces of it to inspect, or even crash their planes into it.

These data, obtained through field studies or by flying high above the continent, and the research stories and histories that glaciologists form around them help people make projections that show up not in formal IPCC documents but in lectures, presentations, and interviews. Through engagements with intimately, intricately extracted data, glaciologists form a sense of the ice that anticipates the behavior of the future ice sheet.

Such anticipation plays out informally in the ethnographic examples described above, as glaciologists and modelers confront the limits of their technoscientific expertise. The field sciences are bound by immediate location, logistics can fail spectacularly, and models need stopgap codes to keep them running. These professional crises are experiences that shape how scientists perceive and inhabit their world, including their interpretations of the future of the ice sheet.

This work is conducted formally in expert elicitation projects, in which research takes experience into consideration in an attempt to make predictions about a world that is not yet understood scientifically. Expert elicitation research affords an opportunity to quantify these felt, subjective, intuitive, and intimate experiences with the ice sheet and compare them with data obtained through models based on standard processes. At least in the case of the WAIS, the predictions that emerge from expert elicitation and the projections that modelers create are noticeably similar and the expert elicitation predictions are more conservative than the objective model data. In this case, experts used their technical and experiential skills to say something provisional about an unknown and uncertain future. Though expert elicitation remains on the edges of acceptable scientific methods because of its inherent subjectivity, it provides an opportunity to give policy makers a little more analytical and predictive information they can use when making climate decisions.

Expert judgment contrasts with popular or public judgment, not only because of epistemological factors but also because of experts' intimate engagement with the world they research. This world is created through long-standing interactions between the researchers and the ice they study, live upon, fall into, and feel saturated by. Though it is difficult to quantify, the experiential domain of scientific practice contributes to expertise. By interacting with the ice and their iterative depictions of it, experts manage to put together particularly informed visions of the future.

Interacting with the nature of one's scientific inquiry is a world-making project in addition to a knowledge-making one. To live in a glaciological world (in contrast to a world populated with glaciers) is to feel and read particular signs; to speak a particular, specialized language; and to create comparisons with distant places and forms. When this happens, science becomes not only a way of knowing but also a way of being, an ontological effort that is relational and practical with regard to the objects of inquiry (Latour 2013). That is, while the disciplining nature of science may be the filter for a scientist's interpretation of the world at hand, the world itself participates by making some relational options easier or more logical and others virtually impossible (Kohn 2013). Sensing the ice is not only an explicitly cultured and disciplined endeavor but also a site where knowledge is formed through embodied experience.

4

SAMPLES AND SPECIMENS AT ANTARCTIC BIOSECURITY BORDERS

The most charismatic and easily identifiable Antarctic life forms are marine species: the penguins, seals, and whales that are found in the Southern Ocean and on the coastal fringes of the continent. They move spectacularly, hunt and attack, and take care of their young. Tourist cruises to Antarctica often emphasize seeking out and photographing these symbols of Antarctic resilience and success.

A bit inland from the coastline, however, there are other flourishing communities of Antarctic life. Found in rocky outcrops, at the tips of mountain ranges otherwise buried in meters of snow, in Antarctic ponds, in the soil, and in ice-free areas, these Antarctic creatures are smaller and usually less showy. Biologists working away from the Antarctic coastline seek out mites, tardigrades, nematodes, springtails, and lichen to study. Antarctic terrestrial biota is are relatively unknown. What species exist? Are they endemic to the Antarctic or can they be found elsewhere? How do they adapt to the severe cold and long periods of darkness? What is the extent of their habitats and how are these changing with the earth's climate and the warming Antarctic? These small creatures, which are almost or entirely invisible to our eyes, are similar to those we could hope to find on other planets and are still relatively mysterious here on Earth. As scientists seek to understand these species, they organize them into taxonomies, ecologies, and categories of threat and endangerment. The discovery of "alien" oceanic life forms, according to Helmreich (2009), challenges scientific understandings: "The figure of the alien materializes, I contend, when uncertainty overtakes scientific confidence about how to fit newly described life forms into existing classifications or taxonomies, when the significance of these life forms for forms of life—and particularly, for secular, civic modes of governance—becomes difficult to determine or predict" (16).

Depending on where Antarctic species are located, they can be positioned as either threatened or threatening. On the continent and in the waters that surround it, indigenous plants and animals are hardy survivors in an extreme environment, symbols of a last wilderness. When these species are taken to other continents, though, they may threaten the species native to that place. They become intruders in an ecosystem. In the context of management practices and conceptual frameworks that fall under the rubric of biosecurity, Antarctic species are contingent species—either interlopers or native residents—depending on which borders the species pass through. The convergence of antiterrorist security programs and invasive species management at the borders means that scientists have the opportunity to influence contemporary security decisions and correct past environmental practices that introduced nonnative species elsewhere. This is a case study of multispecies borders that are made through encounters that involve risk, threat, and the exciting, dangerous, and always generative unknown. All of this is mediated through governance.

The Security of Living Things

This chapter tracks the emergent issue of biosecurity in relation to Antarctic samples and specimens. Biosecurity is an ongoing project that is complex and contentious. While this chapter addresses how scientific ideas about Antarctic species are framed, I am particularly interested in exploring how notions about how to handle these species as they move through the system created by Antarctic policy. These policies have put strict preventative and contingency plans into place in order to ensure biosecurity. To track scientific engagements with Antarctic species, I use some of the geographic terrains that Antarctic scientists and policy makers have acknowledged and constructed through their negotiations of Antarctic biosecurity policy. I will examine these terrains as borders—places where things are allowed through, kept out, contained, and defined by the geographies around them. People who enact biosecurity laws often consider themselves to be "securing borders," although Langlitz and Helmreich (2005) point out that there is never security in biosecurity, only risk. Thus, biosecurity is a set of practices that attempt to manage risk at borders where people, nature, nations, and knowledge are threatened or pose threats.

The borders that are drawn and enacted are good-faith attempts to take action to solve biosecurity problems (for examples of enacted borders, see Bruner 2004). The uses of scientific expertise in the border-making projects below raise questions about how Antarctic scientists can deploy expertise when grappling with environmental problems in relation to their scientific work. The limitations of this expertise, as a form of power to enable effective governance, becomes clear when scientists encounter policy mechanisms that engage the project of biosecurity.

Antarctic biosecurity borders are created from a haphazard and incommensurate jumble of ideas and objects: national claims, a latitudinal line, an ecological convergence zone, the Gressitt Line, thousands of bits of data and their related interpretations and implications, border documents, permits, papers, and policies. These bits of data, printed between commas or as short bullet points, contain legacies of colonial-era Antarctic exploration and claims, fears about the rapidly warming Antarctic Peninsula, and hopes and anxieties about how Antarctic research can affect a nation's knowledge making and ecological integrity and peer-reviewed scientific research.

When biosecurity—a mixture of samples, specimens, paperwork, gear, practices, conventions and policies—is made by one of the borders described below, the scope of scientific knowledge and possibilities shift. The project of biosecurity—an attempt to address a problem that fuses science and policy—illuminates how science is engaged with the Antarctic policy system and how knowledge and expertise shape this policy system. How are organisms contingently positioned as threatened or threatening, as creating the problem or being pummeled by it?

Species matter in the project of biosecurity—their specificity connotes risk and their particular behavior and sociobiology influences biosecurity. Nonhuman species make worlds instead of simply inhabiting ours; the quotidian lives of species intervene in and profoundly reshape the world (Philo and Wilbert 2000; Wolch and Emel 1998). Life is created through multispecies encounters (Haraway 2007; Kirksey and Helmreich 2010; Kirksey 2014; Kohn 2013). Antarctic biosecurity issues—both political and physical—rely upon engagements with species and are shaped by the ways nonhumans thwart anthropocentric management.

The Emergence of Biosecurity

Invasive species, ecologically defined, are species that are not "native" to a place that outcompete and replace native species. I prefer to use the term "nonnative" instead of "invasive," since these species may or may not be invasive. Scientists and managers use the language of warfare when they consider biosecurity problems and the landscapes and beings that constitute them (Martin 1994). The language of biosecurity is almost always militaristic, and alternatives are sometimes difficult to consider when policy makers and scientists frame biosecurity problems as "threats," "invasions," and "incursions." The management of biosecurity problems involves "containment" and "eradication." Describing the species ecologists flag as threatening to specific environments is similarly difficult. The language policy makers and scientists use correlates with the violent, xenophobic discourses and activities that currently surround human immigration issues in the United States and elsewhere.

The linguistic parallels between "native" and "alien" biota and human beings are not always clear: contingent parameters are involved in classifying native and nonnative species, as Helmreich (2005) discusses, using Hawaii as an example. In popular parlance, "canoe species" such as the taro plant that was introduced by Native Hawaiians, are not considered invasive, while species introduced by humans after the arrival of Captain Cook are often viewed as invasive. Scientists and Native Hawaiians dealing with nonnative or invasive species construct situational parameters that make species classifications political and inextricable from cultural and historical experiences. In the case of Hawaii, the dichotomy between an "invaded" landscape and an original, pristine landscape is less clear. A "native" landscape that includes cultivated nonnative species is considered original and pristine. This is a pragmatic landscape, where issues of invasiveness are seen as less important than the work that was undertaken to create a desired place.

The emergence of biosecurity on the international scene is relatively recent and can be traced to several streams. Most directly, state work to protect explicit or implicit "resources" against nonnative species has become increasingly standardized. Frequent and rapid human travel and trade between states and ecosystems creates more opportunities for potentially invasive species to travel to new sites, survive, reproduce, and outcompete native species. Some of the discourses that proliferate about the issue strongly resonate with themes of colonial guilt and a longing to recreate precontact ecosystems in the postcolonial state, as Rosaldo (1989) has pointed out. Government biosecurity programs do not explicitly grapple with these uncomfortable postcolonial political climates in their countries. Instead, they use technical approaches to scientific disciplines to narrate their mission statements and the activities they conduct to enforce strict biosecurity regulations in these states.

Government institutions, media publications, and news broadcasts help promulgate the relationship between nationalism and biosecurity in New Zealand, among other places. In both New Zealand and Antarctica, biosecurity officers are charged with overseeing invasive species problems and implementing precautionary measures at the borders of the nation and internal protected areas, such as offshore islands and national parks. The existence of these officers indicates public support and the government's willingness to make the project of biosecurity a national one. Anyone flying into New Zealand encounters biosecurity checkpoints as they pass through the airport security and customs, where they are required to place fruit and nuts into a biosecurity "amnesty" box or risk fines or jail. In New Zealand, science and security have been pulled together, and nature is managed under a similar security apparatus as terrorism is.[1] Biosecurity, however, magnifies the technologic practices of science in the name of preparedness and a concerned public: ecological disaster and national security are conflated into one domain (Lakoff 2006; Collier and Lakoff 2006).

While biosecurity in Antarctic contexts is not always considered in terms of weaponry or terrorism, it is always constructed in terms of defense, and biosecurity monitoring and responses are always technocratic. These technocratic constructions emerge from tremendous effort by policy makers and scientists. It is important to note, however, that there are alternatives to this defense paradigm—mobile plant and animal life can be imagined, contested, and accommodated in other ways, and in some nonnative species management schemes, these approaches succeed.

As a concept and a set of goals and procedures implemented by state governments, biosecurity is a contested term. Each nation has its own priorities about which biological organisms should be secured. Sometimes different agencies within the same nation have different priorities. As invasive species management, biosecurity has a longer history of governments working to protect ecosystems from invasive threats. In this chapter, I follow several examples of practices at Antarctic biosecurity borders to trace the specific meanings that proliferate among scientists, policy makers, and policy implementers as they try to consolidate what Antarctic biosecurity means, how it might be enacted, and how organisms are contextualized as risk-causing or at risk.

The Technomanagement of Knowledge and Risk

Border making is technomanagement informed by notions of risk and knowledge. Biosecurity borders are tricky in the Antarctic; general Antarctic borders are complicated and contested. The problem with the borders drawn onto the continent is that they are highly disputed. Simply put, very few can agree on where this border is. One can historicize this by looking at the "pie slice" borders drawn on the continent before the Antarctic Treaty, when the original seven claimant states were mapping their Antarctic territories. Most claimants—Argentina, Australia, Chile, France, New Zealand, and the United Kingdom—drew triangle-shaped borders of various widths onto the continent that met in a small point at the South Pole and spread out to include a wider continental shoreline. However, the seventh claimant state, Norway, drew a different border, one that did not extend all the way to the South Pole but wobbles out rather indeterminately into the Southern Ocean (see figure 4).

This differently bordered territory is usually explained in the context of Norway's interests in claiming Antarctic territory to support its whaling industry. But this vague border marks contention over Antarctic borders and their relationship to the continent, the pole, and the ocean that surrounds the continent. The Antarctic Treaty is a document intended to quell national claims to Antarctic territories: it put the seven claims "on ice"—neither accepting nor denying them—and refused to allow any other nations to make claims to territory on the continent (Joyner 1998; Peterson 1988; Quigg 1983; Vicuna 1983).

The Antarctic Treaty member states agree that they make decisions relating to the continent and seas located below the latitude of 60 degrees south. However, some scientists, environmentalists, and policy makers are proponents of a less linear border in the Southern Ocean: that of the Antarctic Convergence, a fluctuating zone about twenty-five miles wide. In this zone, cold Antarctic water sinks below the warmer water from the north. The Antarctic Convergence definitively divides two distinct hydrologic zones, climates, and arrays of life forms. The prevalence of two borders surrounding the continent—a geopolitical line and a shifting line drawn by hydrology—makes it difficult to document the moments when a nonnative species has "crossed" a border.

Thus, the only nonnative species in the Antarctic that can be definitively documented, measured, and controlled are those that arrive from very far away, leaving the closer Antarctic border disputes unsettled. An early example of a nonnative species traveling to the Antarctic during my fieldwork was the North Atlantic spider crab, which appeared in waters near the Antarctic Peninsula (Rincon 2006). These crabs are assumed to have arrived in the Antarctic as passengers in the ballast water of ships traveling between the Arctic and the Antarctic. Both male and female North Atlantic spider crabs have been captured in the Southern Ocean. National Antarctic Program environmental managers call for the elimination of such obvious nonnative species, but in this case, management is virtually impossible. Marine environments do not facilitate easy management and human oversight because they are underwater and in the Antarctic are often frozen over.

The current approach to managing nonnative species is oriented toward prevention, using what is called anticipatory governance or "real-time technology assessment," a process that merges scientific knowledge production with technological innovation (Guston and Sarewitz 2002). Managers feel it is ideal to avoid the arrival of nonnative species in the Antarctic entirely. However, nonnative species are already present in the Antarctic, and it is expected that more and larger invasions will occur in the near future. Thus, some national Antarctic program officials try to continue and enhance current prevention practices and devise plans for managing outbreaks when they occur. It is clear from current practices of nonnative species prevention and management and from concerns about the imminent future that policy makers are trying to comprehend the complex problem of Antarctic biosecurity as scientists are producing information about nonnative species and biogeographic borders. Sometimes policy makers work without any information at all. Despite the existence of precautionary measures and carefully designed plans for managing invasions, great gaps in knowledge exist about which species are native, where niches of native Antarctic species are located and how those species are distributed, or even which species exist in areas of the Antarctic.[2] Policy makers draw biosecurity borders around porous gaps in knowledge, and managers enact policies and practices before the problem is explained, even partially.

Policy advisors describe such uncertainties from the seemingly neutral vantage point of risk assessment, which Jasanoff (1990) describes as "the gray area between science and policy" (216). Jasanoff, following Brooks (1964), claims that there is "science in policy," referring to the fact that policy decisions take scientific expertise into account, and "policy for science," referring to the fact that regulations affect how science is done (5). This duality is noticeable in science and policy work on biosecurity in the Antarctic, where scientific research and policy co-constitute each other. Scientific findings and policy decrees about biosecurity emerge simultaneously and through close consultation with managers, scientists, and policy makers. Scientists often orient the scope of their research to particular risks as determined by "society" or "the world," and their scientific practices thus become attuned to identifying risks and trying to clarify gaps in knowledge. This scientific focus can be useful to policy makers who are attempting to employ the precautionary principle (Beck 1992, 1995).

Antarctic scientists, managers, and policy makers work out their approaches to biosecurity in ways that attempt to resolve uncertainty. Action is taken when the problem gains legitimacy, not when it is deemed "certain" (Douglas and Wildavsky 1982; Jasanoff 1994). In Antarctic biosecurity, science is predictive.

Some small-scale biosecurity measures have already been undertaken in Antarctica. The International Association of Antarctic Tour Operators

(IAATO) has voluntarily adopted practices designed to decrease the likelihood that tourists will carry nonnative species to Antarctic sites. Before each landing, IAATO member tour operators follow standard boot-washing procedures, and they decontaminate tourists' clothing and equipment as well. While these practices are usually framed in terms of environmental ethics, the fact that the IAATO does not mandate them must also be understood as an attempt to resist regulation. By regularly publicizing its environmental standards and practices, the IAATO tries to demonstrate to governments that its members do not need oversight by state governments.

Another example of a small-scale biosecurity practice is the Australian Antarctic Division's response to a multiyear proliferation of mushroom gnats in Casey Station's waste treatment center. In light of Australia's status as a signatory to the Madrid Protocol, the Australians decided that nothing less than total eradication would suffice. The national program relocated all staff from Casey Station and completely flushed the waste treatment center to remove the nonnative species.

These practices demonstrate the kind of attention currently given to nonnative species in the Antarctic environment. Although Antarctic scientists and policy makers working at the biosecurity borders of the continent grapple with significant uncertainty in the realms of scientific knowledge, uncertainties and gaps in research and knowledge do not preclude management practices. International scientists and policy makers collaborate, albeit unevenly, across time, space, and nations to implement practices that attempt to prevent invasions or to eradicate nonnative species before they invade.

Some scientists describe this work as "basic science," a term that encompasses the consolidation of knowledge and observations and the compilation of statistics that help create a snapshot of biosecurity issues in these borderlands. Despite uncertainties about basic delineations (such as the border of the Antarctic region) that makes biosecurity projects somewhat tricky, environment managers try to cope with these uncertainties through ad hoc practices. Scientists draw smaller-scale, ecologically based biosecurity borders on the Antarctic continent and help managers define and mitigate nonnative species invasions. The pragmatic, ad hoc ways that managers respond to biosecurity surprises are the products of a combination of uncertainty about Antarctic ecology and notions about preserving a pristine continent.

Biogeography: Mapping and Bounding Biosecurity

While in the public imagination, Antarctica is often regarded as a uniformly frozen white surface, the continent and its surrounding waters are diverse and can be differentiated into separate zones. When this zoning follows biological indicators, it is called biogeography. Biogeographers concern themselves with the habitat ranges of particular species and how these ranges overlap with other species and move in relation to environmental factors such as surface temperature. Only a handful of Antarctic scientists conduct biogeographical research, and much of their work is currently situated as science in the service of the Antarctic biosecurity problem, primarily in order to obtain funding. Biogeographers are becoming scientific biosecurity experts. At the time of my initial fieldwork, the state of Antarctic biogeography was unknown. But tremendous research and organizational energy has been poured into Antarctic biogeography in the past decade (De Broyer et al. 2014). Scientists, managers, policy makers, and environmentalists have been learning which Antarctic species belong to which regions of the continent and which species are "Antarctic" and which are nonnative. Antarctic policy makers and scientists frequently call for more work on biogeography. Some of this concern arises from the noticeable and dramatic changes recently observed on the Antarctic Peninsula, and the call for further study is beginning to be answered with the drawing of one definitive border on the continent. Categories and borders such as these are often described in academic papers and at scientific conferences, where the individual scientists who make them are situated as biosecurity experts. These forums are where the inscriptions of borders are documented, a practice that has great portent for the future because these scientific documents later become translated into policy texts.

When biogeographers Chown and Convey (2007) compared the species of nematodes, mites, springtails, and tardigrades present at different ice-free areas on the Antarctic Peninsula and on the continent proper, they noticed a significant divergence in species array. Chown and Convey concluded that "indeed, there is a particularly striking lack of overlap at species level between the representatives of several higher order taxonomic groups" (2313). This lack of overlap between the peninsula and continent called for the description of a distinct biogeographic border in the area. Chown and Convey named the Gressitt Line for a pioneering Antarctic biogeographer of the 1960s and 1970s, thus marking and naming a border they have discovered by examining the tiny species that live in small patches of Antarctic ice-free land. Chown and Convey choose to link this border to the emergent issue of biosecurity in the hope that the act of defining the endemic biodiversity of the region would make it possible for scientists to track and manage future nonnative species invasions. This continental border is rooted in the history of the scientific discipline of biogeography and in imagined futures of the Antarctic environment.

The work by Chown and Convey on the Gressitt Line demonstrates some key conceptual frameworks and problems that may emerge in Antarctic biosecurity issues in the future. The research undertaken to describe and categorize native biodiversity in various Antarctic regions is at the cutting edge of Antarctic biosecurity scientific knowledge. Scientists publish this data to lay their information out in publicly available, peer-reviewed

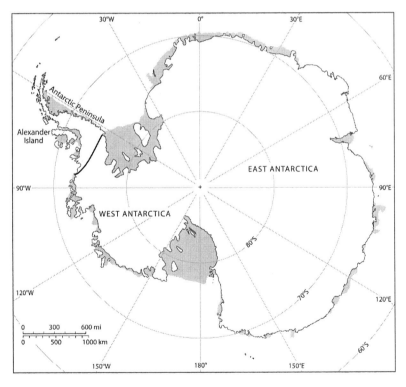

Figure 6. The Gressitt Line. Adapted from Chown and Convey 2007.

formats and to establish themselves as among the few Antarctic biosecurity science experts in the world. By publishing such studies as the Gressitt Line article, Chown and Convey situated themselves as "science advisors" who were available as consultants for the drafting process of other kinds of documents, namely Antarctic biosecurity policy. However, Antarctic scientists are both the sole experts on Antarctic biosecurity and some of the primary sources of nonnative species invasions. Their research is documented not only in academic journals and conferences but also in mundane negotiations that take place at other borders. This work takes on a significant valence of nationalism, particularly in relation to security, when we examine what happens when Antarctic native species become nonnative—that is, when scientists carry their Antarctic samples and specimens to other places.

Antarctic Samples at the New Zealand Biosecurity Border

The border between New Zealand and Antarctica is enacted primarily at the Christchurch International Airport on New Zealand's South Island. Those who bring Antarctic samples and specimens into New Zealand must secure permits to prevent biosecurity breaches. Rocks, water vials, and living and dead Antarctic bodies could introduce unknown organisms into New Zealand, a nation with a high number of endemic species and a history of extinction and threat caused by invasive species. Many people in New Zealand, including the scientists I depict below, support biosecurity laws and are willing to comply with strict rules at the airport and in other places that are potentially sensitive to the introduction of nonnative species, such as heavily visited trailheads in national parks. At the trailheads, walkers stand in pans of disinfectant to kill unwanted organisms from their boots before starting their hikes. While Antarctic environmental managers and scientists are similarly concerned about the possibility that invasive species could colonize the Antarctic continent and surrounding oceans, this section will focus on the arrival and movement of Antarctic biological samples in New Zealand.

The biosecurity border between New Zealand and Antarctica is a zone of anxiety where the tension between contamination and discovery is played out as Antarctic samples arrive. On one hand, the origin of these samples outside New Zealand's national borders signifies danger and a need for meticulous regulation, particularly for a nation whose economy relies on agriculture and nature tourism. Tourists arrive in New Zealand expecting a pristine and magical "Middle Earth,"[3] and the official national tourism board advertises a "100% Pure New Zealand." Nonnative species foment intense concern in relation to this projected image. On the other hand, Antarctic samples at the biosecurity border represent the frontier of New Zealand's scientific knowledge production and are full of the promise of new discoveries. Antarctic samples are both aliens arriving from beyond the nation and, in part, what gives power to technoscientific knowledge in the nation. The samples may help scientists produce new knowledge about nature, a project that states support and that gives them power.

Because their regulations allow or disallow various scientific practices, national science policy makers interpret and shape new scientific destinies (see, for example, Abraham 1998; Anderson 2002; Mukerji 1990). In a study on policies designed to reduce gorse in New Zealand, Barker (2008) notes that "biosecurity practices in New Zealand produce a complexity of semi-permeable control boundaries that are flexible and sensitive to the shifting spatiotemporal geographies of indeterminate entities, and to changing and competing human values" (1612). New Zealand's flexible approach to biosecurity is seen as ideal because of its ability to deal with problems as they arise (instead of before or after they occur). It also produces a sense of continual negotiation among officials, people carrying organisms that relate to biosecurity, and the species themselves. In the cases below, samples, scientists, and officials from the Ministry of Agriculture and Forestry (MAF)[4] gather at the New Zealand-Antarctic biosecurity border, and address the promise and mitigate the anxiety wrapped up in small containers of Antarctic samples and specimens at the New Zealand border.

Much of the biosecurity border work occurs at the MAF's (now the Ministry for Primary Industry's) biosecurity checkpoint at the Christchurch International Airport. The extensive and complicated permitting system for samples and specimens sorts them into different secure facilities as they arrive from the Antarctic. Antarctic scientists from New Zealand secure permits through the ministry before their departure south and then must pass through customs at the international terminal in Christchurch when they return, holding both their permits and their collected samples for inspection by MAF officials. The scientists, still kitted out in the extreme cold weather gear they are required to wear on flights to and from Antarctica, stand in line alongside businesspeople flying from Japan or Korea and tourist families returning from Fiji. In their layers of bulky down, thick fleece, and overstuffed blue mukluk boots, the Antarctic travelers attract attention from the others waiting in the customs line. Each person holding samples must provide detailed information about what is contained in the samples, where the samples will travel to in New Zealand, and what else could be contained in association with the samples. A successful pass through the biosecurity checkpoint entails obtaining clearance to move the samples into an appropriate, permitted facility in New Zealand. For Antarctic scientists, this usually means storing their samples or specimens in specific rooms in or near their laboratories. The ministry scrutinizes these rooms and their storage facilities, ventilation systems, and access points to prevent the introduction of nonnative biota into New Zealand.

Occasionally, this sorting and storing process gets jammed up between the government's two projects of science and security. I met a scientist named Gertrude at the worm lab of her university, where worms were stored, frozen, and studied. Not surprisingly, the lab had an earthy smell, vaguely reminiscent of sidewalks after a rainstorm. I asked her about her experiences with the Ministry of Agriculture and Forestry. She said that she arrived at the MAF airport biosecurity checkpoint from her fieldwork in the Antarctic toting two coolers packed with frozen soil samples that contained nematodes, microscopic Antarctic worms. Gertrude's university department had a general permit for bringing specimens and samples from outside New Zealand into its laboratory, and her copy of the permit had arrived shortly before she headed to Antarctica. When she re-entered New Zealand, though, an MAF official discovered an error on the permit that stated that the Antarctic samples had originated in Israel. The official would not allow the samples through the border, and they were stored in the Antarctica New Zealand airport freezer for weeks, next to the program's steaks and pizzas, until the paperwork error was eventually corrected and the samples could be transported to the secure university laboratory. The nematodes in this story were doubly threatening because they were a nonnative species and because their paperwork was obviously out of order. Because of anxiety about the origin of the samples at the biosecurity border, MAF officials blocked the samples' passage across the border. This threatened the lives of the nematodes because of potentially inadequate storage. The specimens became insecure at the border, and the border also became insecure through its housing of the nematodes. Both the safeguarding of the nematodes, which hold the potential to contain scientific knowledge, and the safeguarding of the border, were compromised in the anxiety-filled zone of the biosecurity checkpoint.

Another example of Antarctic samples troubling the biosecurity border involved my friend Demelza and her several liters of pond water samples. Demelza and the samples got through the Christchurch airport biosecurity checkpoint with little problem. Her name differed from the main name on the permit, but Demelza pointed out that the permit was a group permit, not an individual one. After the officials made a few phone calls and constructed a detailed written record of the pond water samples, Demelza was allowed to drive her samples to a laboratory facility in Christchurch.

The problems began when Demelza called the MAF to get a permit (called an "authorized movement permit") to transport the pond water to Auckland, where the equipment she needed to analyze the chemistry of the water was housed. The MAF refused her permit application because it turned out that the laboratory in Auckland was what is called a "transitional facility," where samples are inspected, tested so they can be categorized as safe or not for the New Zealand environment, and then destroyed. Since Demelza wanted to keep some of her pond water in case she needed to conduct further research, a transitional facility was not appropriate, and the MAF refused to authorize such a movement. A flurry of e-mails between Demelza, her advisor, and the MAF took place for the entire year that I lived in Christchurch, and a new potential solution was always on the brink of solving everyone's problems. The MAF and Demelza tried to figure out a way to remove the biological components from the water samples without compromising their chemistry: in official language, the MAF asked Demelza to "devitalize" the samples. However, there is no clear-cut definition of how to devitalize water, and the MAF and Demelza debated about which methods were appropriate. First, the MAF suggested that she acidify the water. She rejected this, since adding acid to water of course alters its chemical composition. Then they asked her to run the samples through an autoclave, which involves heating the water to a high temperature. Demelza also refused this suggestion because the high heat could alter the chemistry and superheating Antarctic water samples would look dodgy in the methodology section of her dissertation. If there were biological components in her water sample, there was a chance that these components contributed to the chemistry of the pond. While both devitalization methods would have killed off any potential microorganisms, Demelza argued that they would have fundamentally changed the samples to the point that were no longer Antarctic samples. The proposed devitalization of the water did not overly concern Demelza, but the means of achieving devitalized samples possibly would have destroyed other components of the water that were integral to her research.

Demelza countered with a suggestion that she treat the water with gamma radiation. The MAF refused this solution since the samples would have had to have traveled to Wellington to be radiated, making an unauthorized movement. Demelza's final suggestion was that the samples be treated with ultraviolet rays, which could have been done at the Christchurch lab where her samples were stored. As I left New Zealand, she was waiting to hear back from the MAF about this suggestion. She was concerned because the MAF requires 100 percent devitalization and UV treatment offers only 99.99 percent devitalization. Her samples had been stored for ten months without being analyzed.

At this New Zealand biosecurity border, it is assumed that the samples are alive and the ministry required them to be rendered not alive. When that takes place, the danger lies in the samples being compromised to the extent that the anticipated scientific knowledge cannot credibly be read from them (on reading knowledge from sampled bits of things, see Cruikshank 2001; Haraway 1989; and Shapin 1994). Also, officials from the MAF (now the Ministry for Primary Industry) were not the only ones who conducted the government's work of giving the samples clearance to travel around New Zealand. In this case, as an expert scientist, Demelza participated in this work too, refusing certain devitalization techniques and offering suggestions that were more appropriate for the specific scientific possibilities her Antarctic water contained.

The tension at the biosecurity border is produced by the different kinds of work done by the scientists and the government employees who work at the airport checkpoint and who respond to e-mails and issue permits. The scientific and bureaucratic work done on both sides of the border is wrapped up in the work of the nation. People involved in conflicts between the government's projects of science and security try to work out solutions that, at their best, will ensure that no nonnative species will affect the New

Zealand environment and preserve the knowledge-making capabilities of the samples as scientific subjects. The relationship between science and security here is one of co-dependence: security underwrites science and science underwrites security, each making the other.

This border's human and other biological interlocutors trouble the goals of biosecurity. The work government officials do with scientists toting samples from the Antarctic is bound up in national environmentalist ideals and specific techniques of governing science and scientists. These techniques are bound up with and complicit with the expertise of the scientists and their ability to read data and make knowledge from their samples and specimens, which carry both risk and promise.

The Borders, Interfaces, and Crossings of Science and Policy

The emerging issue of Antarctic biosecurity illustrates the relationships between science and policy at various biosecurity borders. Antarctic policy makers and scientists primarily see biosecurity as a management issue. However, managers rely on political and scientific expertise to construct their strategies for responding to this issue. The different Antarctic biosecurity borders reflect consideration of these contributions from scientific experts, and expert recommendations mingle with pragmatic, possible management activities at all scales of the biosecurity problem with varying degrees of success.

As both a scientific concept and a management goal, biosecurity is both made possible and thwarted by the Antarctic policy system. As it travels, it picks up particular valences and discards others. Antarctic samples and specimens wrapped up in biosecurity problems illustrate two concluding points. I cannot ignore the fact that my study of biosecurity borders also became an analysis of documentation—perhaps the most mundane but most important pairing in any border studies. Even borders that are not state lines—for example, the biogeographical Gressitt Line—are essentially comprised of and justified by a flurry of powerful documents. These documents are the materials that form the border. I have intentionally resisted conflating nonnative species issues into issues of national security and the language of aliens, but is it possible to resist the trope of documentation at borders of any sort?

Also, biosecurity makes the contingency of species possible. While species often are situated in the apparently fixed and complex web of taxonomic systems, in the Antarctic context, how species relate to each other and their environment depends on the specific situations at hand. The meanings and properties of these beings shift as they are passed through various Antarctic borders. Such contingencies are not accounted for in Antarctic science and policy, and that is what produces the problems described above.

Working on the problem of Antarctic biosecurity also reveals the threats and promises of Antarctic science and policy to a broader degree. The citizenship of an Antarctic scientist matters, even though the scientist is also entrenched in the international project of Western science and is encouraged in the Antarctic Treaty System to cooperate and share information. But the movement of scientists and their data can be seen as a threat to both Antarctic environments and those to which the scientist returns. While Antarctic scientists can pose risks to the environment, they can also pose risks to policy as they mobilize their expertise to actively challenge national bureaucrats about the policies that hinder their work.

5

MANAGING ANTARCTIC SCIENCE IN AN EPISTEMIC TECHNOCRACY

Turning what one knows about the physical and biological nature of a place into management strategies that take into account geopolitics, cultural concerns, and legal structures requires a suite of techniques that incorporate technical, scientific, and bureaucratic and legal approaches. Those who attend to matters at the frontier of science and policy often situate science as the primary font of information and expect that policy will emerge from that source. Critiques of this endeavor usually point to the impartial and unsatisfactory ways that science manages to show up in policy.

Instead of dismissing how policy fails to reflect science, I will take that failure as a given. My focus is on how pieces of scientific research travel into policy, particularly in how a science-oriented policy apparatus refers to science not just in policy text but also in the form and matter of policy institutions and practices. That is, science and its conceptual underpinnings shape policy practices. The form and matter of scientific practice is essential in creating the bureaucratic and technical art of policy making.

Researchers plan an academic life for their research findings, which they publish in peer-reviewed journals and present to their colleagues at meetings. But to a state, science has multiple uses: otherwise, why would so many national governments be pouring millions of dollars and thousands of hours of military logistical support into it? This chapter explores one trajectory of scientific research into international policy. To conduct this exploration, I will continue tracking Antarctic biosecurity. Antarctic policy makers are concerned about the possibility that nonnative species will be introduced into the Antarctic and consider the combination of increasing human visitation and the fact that some Antarctic regions are warming rapidly to be accelerating the biosecurity threat in Antarctica.

This problem cannot be managed without scientific information. But not every Antarctic policy maker can become an expert on biosecurity science. Which scientific articles, case studies, and species manage to travel through the Antarctic Treaty System, all the way through to consideration at the international Antarctic Treaty consultative meetings? Why do some bits make it while others do not? How is scientific information talked about and treated at high levels of international diplomacy? Policy decisions are often framed to be based on the availability and consideration of "sound science": in this chapter, I show how science is distilled to become "sound" and relevant to policy.

This chapter explores the techniques people use to make policy, in a material sense. In particular, I will dissect the concept of epistemic technocracy to reveal how policy making is technical, bureaucratic, and driven by expertise. To do this, I will track moments when scientific ideas come to matter in the construction of environmental policy. This occurs at the intersection of bureaucracy and technology, where a terrain is formed that is technocratic by design. Epistemic technocracy is also material, as it relies on the discursive forms and particular objects of scientific practice, when explicitly handling scientific data, but also when science knowledge is peripheral to governance.

Governance at the Intersection of Bureaucracy and Technology: Scoping Techniques

Meetings of experts have a particular form. Participants carve time out for them and agree to be present. These meetings are also constitute heavily ritualized, mediated conversations. Expert meetings are one-off workshops meant to focus attention and expand upon a suite of ideas about a particular topic. That experts are invited and attend suggest that a pooling of expertise is brought to bear upon the topic. But this does not always happen easily or elegantly. The negotiation of expert ideas into a policy-relevant document requires techniques that are sometimes pedantic and awkward and sometimes mimic scientific practice.

Since scientists are the primary spokespeople in defining the emerging issue of Antarctic biosecurity and are among the people most likely to cause or prevent nonnative species invasions, several of them were invited in 2006 to the Non-Native Species in the Antarctic workshop at the University of Canterbury in Christchurch to collaborate on creating international Antarctic biosecurity policy. The university- and government-based organizers invited scientific and legal experts to help shape policy ideas. These experts pooled their ideas in a draft policy document that was intended to encompass international Antarctic biosecurity. The workshop was an intensive moment for defining the issue of Antarctic biosecurity, particularly because scientific uncertainties were paired with a sense of urgency about creating effective policy and management practices to respond to this issue. The workshop was small; only about forty invited participants attended. In addition to putting together a paper to present at the upcoming Antarctic Treaty Consultative Meeting, the event also solidified New Zealand's role as a leader on Antarctic biosecurity concerns. Henry, the environmental manager for Antarctica New Zealand, was one of the leaders of the workshop: his role as a delegate to the international Antarctic Treaty Consultative Meeting meant that he would work on the topic of the workshop in that forum as well.[1] This event offered a moment when experts could broaden an emerging policy issue in a style that was less constraining than the typical policy work of formal meetings and Internet-based work in between the meetings.

That this workshop existed at all is somewhat fortuitous. Most policy papers in the Antarctic Treaty System are put together by one nation or are created internationally through electronic communication. However, Antarctica New Zealand had received a donation from the producers of *Happy Feet* (dir. George Miller, Warren Coleman, and Judy Morris, 2006), an animated song-and-dance romp featuring tap-dancing penguins and Lovelace, a guru-like penguin with a plastic six-pack ring around his neck, voiced by Robin Williams. The producers of *Happy Feet* wanted to support management of the Antarctic environment, so Antarctica New

Zealand received an unexpected cache of money that it funneled into the Non-Native Species in the Antarctic workshop. To recognize the producers' contribution, the workshop organizers screened *Happy Feet* trailers at the end of the first day of the conference, causing some laughter and shifting in seats among the workshop participants. This sponsorship is important to note because it demonstrates the contingency of Antarctic policy making and the financial clout of the continent's most charismatic residents. Without the tap-dancing animated penguins, the serious work of starting to create Antarctic biosecurity policies could not have taken place in this form.

Representatives from several nations were present at the workshop, including New Zealand, the United States, Australia, France, and Italy. The participants had professional roles in academia, governments, and NGOs. Many knew each other from similar meetings, and the tea breaks and discussions were friendly and joking. The first day was devoted to lectures from an international cadre of biosecurity experts. The second day began with more lectures, but then the workshop participants were split up to conduct their work for the next two days. The workshop organizers formed working groups in which participants represented various "stakes." These groups filtered into assigned rooms to complete their tasks.

Through my affiliation with the University of Canterbury's Gateway Antarctica department, I was asked to be a note taker for one of the working groups. To facilitate this task, I loaded a PowerPoint presentation designed by the workshop organizers onto my laptop. The presentation consisted of headings only, with the content portion of the slides left blank. The blank outline asked the members of the working group to organize their ideas into the following topics:

1. Prevention of introductions *into the Antarctic* (from outside Antarctica)
2. Prevention of spread within the Antarctic once non-native organisms have arrived
3. Prevention of introduction of species that are native in the Antarctic *to other areas in the Antarctic* where they are not native
4. Prevention of introduction of Antarctic Species into other areas (that is, north of the Antarctic Treaty area) where they are not native
5. Protection of Antarctic values against non-native species impacts: what are we trying to achieve?
6. Legal/institutional issues

7. Science and research
8. Future considerations
9. Other

The working groups were asked to fill in the content of the slides based on their conversations. The initial slides sketched out several boundary-making projects: invasions of nonnative species from outside the Antarctic continent and invasions of Antarctic species to elsewhere. The instructions for the working groups encouraged participants to think in terms of gaps, pathways, and taxa in their considerations of topics. Workshop organizers encouraged participants to think about gaps in this way:

> When identifying "gaps," include procedural/regulatory, knowledge, awareness, technical & practical aspects etc. as you think appropriate. Give some thought to as many pathways as possible, and all types of taxa. . . .
> When identifying ways to address "gaps," include, as you think appropriate

- What can be done at a practical (operational) level (by operators, by individuals etc.)
- Can this be done with existing regulations? If so, how?
- If it would need additional regulatory development—what would it be (regulations, codes of conduct, standard operation procedures? . . .) and at what level (ATCM, SCAR, CCAMLR, national operators, NGOs, others, etc.)
- Significant impediments to mak[ing] this happen, (e.g., lack of awareness, lack of knowledge, . . .) and how these impediments could be overcome.

Working group members were instructed to fill the blank PowerPoint slides with lists and considerations of the multiplicity of gaps in relation to several Antarctic biosecurity borders. This is similar to Collier and Lakoff's (2006) preparedness activity, by which security experts brainstorm about potential scenarios and try to organize potential responses. At the workshop, the generative and at times contested practice of filling out forms was mediated by the structure of the PowerPoint slides. The structure of this specific form produces affordances for the kinds of collaborations and results that may emerge in the final product (Sellen and Harper 2002, 16).

The leader of my group was Samantha, the environmental manager for a national Antarctic program. The participants in the working group

I was in introduced themselves; named the agencies, organizations, and universities they worked for; and identified their primary research or policy interests in the Antarctic. As the conversation moved to discussing each point on the PowerPoint slides, a few group members stood out as particularly outspoken. Charles, a biogeographer, and Steven, a graduate student conducting a study of hull fouling, repeatedly emphasized their own research, especially when Question #5 was discussed, which charged the group to come up with a list of key concerns. Charles, who had given a scientific presentation during the lecture portion of the workshop, always jumped in to emphasize the need to distinguish between the biogeographical zones, the topic of his current research project. Steven prioritized hull fouling, ballast water, and marine issues repeatedly and to the exclusion of any other topic. He had been chosen to represent marine Antarctic issues, and he did that to a fault. I typed out the ideas the working group produced. At the end of the day, Samantha asked me to read the lists for each point out to the working group. As a group, participants decided which points were critical. Samantha asked me to highlight these points. From these whittled-down lists, Samantha encouraged the group to choose a few points that were particularly important or innovative. She asked me to highlight these in a different color. For Question #5, one list looked like this after the highlighting procedure:

- What are we aiming to achieve? Are we only aiming at avoiding species extinction, or is there more?

1. Preventing extinction
2. Prevent establishment (preventing eradication)
3. Barriers for the Antarctic should be at higher bar: Antarctic ecosystems in more pristine state/ perfectly pristine state: important to protect that intrinsic worth
4. No introductions (impossible but should be the intent)
5. Preserving subspecies and genetic diversity (at least minimizing anthropogenic effects—not natural migratory patterns)
6. Scientific value (ex.: boundary between peninsula and continent). Impact in terms of research resource
7. Reciprocal transplants issue: now worried about movements of organisms (scientific community to blame for the current Antarctic introductions)

Almost all of the items on this list of the primary goals of Antarctic biosecurity were highlighted as critical. The most important point on the list was one of Charles's contributions: preserving subspecies and genetic diversity.

Other group participants contributed in less direct ways: their ideas were often bracketed in parentheses and were not boldly highlighted at the end of the day. A group member named Mimi had been invited to the workshop to represent the positions of environmentalist NGOs. She is a firm advocate of including NGOs in the Antarctic policy-making process, primarily because of the "additional brains to pick" they provide. She sees creative and environment-oriented ideas as the most useful contribution environmentalist NGOs can make in Antarctic management. Mimi often presented ideas outside the parameters of the rest of the group's discussion, but she always characterized them as "out of left field."

Charles tended to dominate the conversation, although Samantha would eventually steer the group back onto track. Charles, outspoken, frank, and good-naturedly gruff, did not just comment on issues relating to his own research; he also frequently commented on the state of international Antarctic politics. At one point, he spoke about the Antarctic Treaty System generally, stating that he is not convinced of the ability of the system to do anything and that he does not think the Antarctic Treaty has "any teeth." Mimi responded to Charles by stating that this is simply the nature of international conventions, that each state enforces its own activities, and that such conventions rely on goodwill, "peer pressure," and "embarrassment."

At the beginning of the session, the group got into some complex discussions, particularly relating to the "wilderness value" the Antarctic Treaty tries to protect.[2] One person remarked that the wilderness value in Antarctica is more "extreme": less infringement is tolerated in Antarctica than other places and people are expected to protect more in Antarctica than they are in other places. Someone else noted that economic values in Antarctica (implying tourism) are linked to the preservation of the wilderness values. Another participant tried to steer the conversation away from using the word "wilderness" toward something like "natural" because that participant felt that the concept of wilderness inherently implies association with human activities and experiences. There seemed to be consensus among the group that Antarctica had several "values" to be protected in relation to nonnative species. Wilderness and science values were emphasized as was what is acceptable and expected vis-à-vis protecting these values.

The mobilization of values as concepts allows for epistemic and technocratic work. As I explained in chapter 3, the language of the Madrid Protocol suggests that policy makers seek to protect values rather than other concepts, such as land, water, or natural resources. Because of this, Antarctic environmental managers often hinge their conversations on the protection of wilderness values, aesthetic values, or scientific values. This approach allows for definitional work that permits participants to work toward consensus about these values through conversation. Centering deliberations on the management of values rather than on protecting more material targets is also a bureaucratic device that makes it possible to enfold epistemological difference into broad categories without having to settle them. Values conversations can be rich and can be a way to approach consensus.

During the working groups, some participants shared examples of what worked for their programs and agencies and examples of Antarctic and sub-Antarctic biosecurity problems and how they had been dealt with. These anecdotes were listened to with interest, partially because there is no consistent reporting of nonnative species arrivals among national Antarctic programs. Antarctic nonnative species events are not always found in scientific peer-reviewed literature because they usually occur at bases and it is base managers and employees—not research scientists—who usually deal with problems. Although such incidents might be described in a base newsletter or in another informal internal document, they would not be written up in scientific journals. So one of the working group's suggestions was to improve and speed up information sharing as it relates to nonnative species and die-offs in Antarctic places. That biosecurity information travels through infrastructure networks instead of scientific ones marks a moment of rupture when biosecurity policy fails to follow scientific practice, although most of the time people work strenuously to align policy with scientific understandings.

People not as strongly connected to Antarctic programs, such as those who were from Biosecurity New Zealand or New Zealand's Department of Conservation, shared stories from their agencies that involved invasive and alien species strategies that might be transferable to Antarctica. One of the major recommendations of this working group was the idea of "rapid response frameworks"—a compilation of papers, reports, and contact information that would represent a varied set of institutional knowledge about how people have handled invasive and alien species events in Antarctic and non-Antarctic places. With all of this information in one place, the group thought, managers and other decision makers could more easily respond to an outbreak or an invasion in a well-informed way. Several group members repeatedly stressed that responses should be species (or taxa) specific and potentially specific to the biogeographical regions Charles had outlined.

Mimi stated that preliminary environmental evaluations (PEE) might be one place where it would be easy to start scrutinizing plans for potential invasions of nonnative species. She suggested that using the paperwork system already in place would, at a minimum, encourage scientists traveling south to think about nonnative species in relation to their research and logistics. Some people in the group thought that this would be "just another box to tick" when going through the pre-departure paperwork. Others thought that even if it was just a box to tick, it would still show up in some small way on people's radar on a regular basis. Others thought that the evaluators of these PEEs could take these seriously and keep sending them back unapproved until the preparer demonstrated significant thought about and plans for preventing invasions of nonnative species. To this working group, paperwork could become a tool for changing thoughts and behaviors. The history of environmental impact assessment as a type of potentially powerful paperwork was salient, familiar, and promising to some people in the group.

On the third day of the workshop, the entire group met together to present the findings of each working group. The presentations were light-hearted and the participants joked amid the serious policy recommendations each group made. Henry, the environmental manager for Antarctica New Zealand, summed up the "quick wins" that were gained from the workshop and thanked the participants for attending. He urged participants to communicate the outcomes of this workshop. Over the next couple months, Henry and his staff compiled and condensed the material from the workshop discussions and the filled-in PowerPoint presentations into a working paper to submit to the Committee for Environmental Protection at the Antarctic Treaty Consultative Meeting in Edinburgh.

The Non-Native Species in the Antarctic workshop was a brief but intense moment when different types of experts compiled their ideas onto documents and sorted out which anecdotes and policies looked like the "best practice" in specific biosecurity scenarios. Sometimes more outspoken experts were able to make their ideas more visible in the resulting documents, but the practice of editing and debating about the content of the PowerPoint templates also made possible moments of rough consensus. The collegial environment of the workshop fostered such consensus-making. The ideas, lists, and policy suggestions the workshop produced could be whittled down and retooled to be presented in a formal paper at the Antarctic Treaty Consultative Meeting.

The technical practices of biosecurity, at this moment, were ordered into a type of document called a scoping document. Such documents are a point of entry for policy makers and create space for a policy conversation. These documents outline the breadth and depth of a particular topic without explicitly recommending a particular policy. The scope, however, is presented in particular ways that leaves other points unsaid, undescribed, and off the table. Some participants at the workshop were interested in broadening the scope of the conversation while others were interested in politicking. While this chapter often returns back to the papers the workshop produced, I want to emphasize that the papers are material traces of conversations that are ordered in particular ways to underscore and obstruct particular lines of thought, even though they are meant to be generative, to allow a space for idea sharing that is less constrictive than spaces explicitly constructed to produce policy recommendations. Most seasoned participants anticipate broad discussions, conversations that veer temporarily off track, and attempts to push thinking and policy strategies a bit. The skills and attention poured into this document differed from those that produce the elements of more formal policy documents.

Narrowing the Scope: Formalizing Policy Maneuvers and Making Papers Work

At the level of international Antarctic government, documents—that is, policy papers—are the simply material objects that may or may not translate into international, consensus-based decisions. In formal meetings that use diplomatic forms of communication and a regimented set of procedural rules, it is expected (or at least hoped) that the documents will provide action-oriented policy work. Policy work can range from meeting a procedural obligation to provide information to the other consultative parties to a recommendation for a binding measure agreed upon by the members and taken back to state governments for adoption. This section tracks how science is put to work in working papers. Expertise here is performed differently than it is when scoping documents are produced. At the ATCM, the term "expertise" is used as technocratic shorthand for the actions of bundling research complexities, scientific assumptions, and conflict into concepts that might bring about consensus.

An Antarctic Treaty Consultative Meeting consists of two weeks each year when the members of the international governmental regime for Antarctica meet to share information, get plans approved, and make decisions.[3] The entire plenary meets rarely; most of the two weeks are devoted to smaller groups that report back to the plenary. One of the most active groups is the Committee for Environmental Protection, which meets for most of the first week of the ATCM. The CEP is the group that oversees the Madrid Protocol (1998), the environmental policy document for Antarctica. The ATCM is a large, formal attempt to consolidate all the work completed during the year, including the work done in workshops and intersessional contact groups, whose members discuss issues in online forums.

At the 2006 meetings of the ACTM in Edinburgh, the CEP met in a large room at the conference center. CEP state representatives sat at tables arranged in a large rectangle. At each seat, delegates found their microphones, translating headsets, the flag of their country and a sign with the country name, glasses, bottles of water, ATCM paper pads, pencils, and bowls of chocolates and mints. The state delegations sat in alphabetical order. The tables were draped in blue, and the head of the table, where the chair, his assistant, and an Antarctic Treaty Secretariat representative sat, was covered in "Antarctic tartan," a nod to the Scottish location of that year's meeting. The tartan pattern had been designed in honor of that year's Antarctic Treaty meetings, and each delegate received a scarf in this pattern with their meeting materials when they registered. A row of flags stood at the far side from the entrances, one for each of the CEP members. On the opposite side, behind the member state tables, was a back row of tables. Representatives of Antarctic Treaty signatory states who had not yet signed the Madrid Protocol sat at one end of the back row. Next, the Malaysian delegate sat by himself at one solitary station—at this point, the Malaysian government had refused to sign the Antarctic Treaty but had been invited to the ATCM to encourage it to consider signing.[4] Observers and experts filled the remainder of the back row: representatives from the Antarctic and Southern Ocean Coalition, the International Association of Antarctic Tour Operators, the Scientific Committee on Antarctic Research, and so on. The tables in the back row had everything on them that the main table did except the mints and chocolates. Each of four booths across the room from the head of the main table housed two translators who translated into the four official languages of the ATCM: French, Spanish, Russian, and English. Large television screens projected Power-Point presentations or images of speaking delegates. Behind the ASOC seats, stacks of photocopied information papers and secretariat papers were available in all of the languages. All of the working papers could be found behind the head of the table.[5]

During the meetings, there was often whispering among the delegations. There was movement as a delegate would go over to speak to another delegation about something as the formal proceedings carried on. Delegates passed around papers, edited them, and returned them in a quiet, multitasking negotiation of text that had been requested by the chair on particularly contentious parts of the final report. As the languages used shifted, delegates put on and took off their translation headphones. Some people kept their headphones on at all times, as their channel continuously transmitted in their selected language.

The chairperson (an Australian) or his aide made sure the meeting followed its agenda. The delegates introduced and debated working papers (which can translate to actual policy measures) and information papers (which cannot contain measures). Only member states and officially designated observers can submit working papers. Organizations such as ASOC that are designated as expert organizations can supply only information papers. Some papers invited congratulations from the table; some inspired conflict, defense, or silence.

How these papers are received is not left to the whims of the delegates at the moment of formal presentation. Authors of papers circulate their ideas and drafts to other delegates, coordinating a formal response from the sidelines of the meeting. For example, authors from the ASOC lobby national delegates to provide positive comments of support. These comments legitimize some of the environmental work ASOC proposes. Importantly, interventions become part of the report: while a paper with no comments occupies only one line of the final report, a chain of comments in the final report indicates interest, support, or conflict—in short, engagement with the ideas presented in the paper. Delegates can indicate these moments of productive conversation in the aftermath of the meetings as they report back to their governments, members, or supporters. Thus, these informal sideline negotiations require technocratic savvy as individuals work to harness interest in and attention for ideas presented at the ATCM.

The delegation from New Zealand presented the working paper on nonnative species that emerged from the workshop in Christchurch. The paper received unequivocal praise and numerous statements of support from the state delegations.

In the final report issued by the ATCM, the working paper on the biosecurity workshop was mentioned by the United States. The CEP report on the meeting describes the comments in slightly more detail. It reads:

> (127) New Zealand introduced WP 13 Non-native Species in the Antarctic. Report of a Workshop and the accompanying IP 46 "Non-native Species in the Antarctic:" a Workshop. The workshop took place in New Zealand from 10 to 12 April 2006 and arose from discussion at CEP VIII of Australia's ATCM XXVIII—WP 28. The key issues outlined in the paper included, but were not limited to, the transfer of species both into and within the Antarctic and the need for practical preventive measures.
>
> (128) Several Members thanked New Zealand for the paper and agreed that the introduction of non-native species to the region requires close consideration, particularly as a warming climate is expected to increase the ability of new introduced species to survive in the Antarctic. The Chair recalled that similar issues had been recognised as important during the recent CEP Workshop on "Antarctica's Future Environmental Challenges" (WP 42).
>
> (129) The Committee strongly supported the six key recommendations arising from the workshop, including that:
>
> - the issue of non-native species should be given the highest priority consistent with the high environmental standards set out in the Protocol; a "zero tolerance approach";

- the CEP should take the lead on this issue;
- the CEP should give consideration to sharing information with, and seeking advice from, other bodies, notably SCAR, CCAMLR, COM-NAP, IAATO, IUCN and other organisations as appropriate (e.g. IMO);
- dedicated research is required to improve understanding of, inter alia, existing biological and genetic diversity, species distributions and biogeographic zones, the potential implications of a warming climate and identification of high risk areas and ecosystems; particular research attention needs to be given to microbial communities and marine ecosystems;
- to the extent possible, non-native species issues concerns should be built into existing procedures and practices notably EIA procedures and the protected areas system; and
- a set of comprehensive and standardised guidance and/or procedures should be developed, aimed at all operators in the Antarctic, based on a "Prevention, Surveillance, Response" approach (ATCM 2006, 287–288).

These points include allusions to surveillance and preparedness and calls for future scientific research. Several of these points directly referred to some of those written on the PowerPoint slides of Samantha's group at the Non-Native Species in the Antarctic workshop. In particular, Charles's calls for bioregionalization, especially for the microbes that are his specialty, were flagged for future research. Steven, the marine interest stakeholder from Samantha's group, also got his ideas into this short list of research needs.

However, the issue of biosecurity became a small blip in the overall discursive fabric of the Committee for Environmental Protection meetings, which covers just a portion of the mass of information presented, negotiations conducted, and decisions made at the Antarctic Treaty Consultative Meeting. Nevertheless, the New Zealand working paper achieved the work its authors desired: a new topic was introduced to the international arena and several national Antarctic programs are now doing some sort of work on the issue. The fact that the biosecurity issue was brought up and well received at the CEP was the result of years of relaxed and friendly dialogue that may at some point in the future result in the passage of a biosecurity measure after many states have independently implemented biosecurity procedures in Antarctica.

A year later, at the 2007 ATCM in New Delhi, India, the discursive climate of the CEP had shifted somewhat. First, the U.S. delegation had been cleared to acknowledge the scientific reality of climate change, which the delegates did enthusiastically at each possible opportunity. This shifted the tone of the meetings overall. Henry, the former chair of the Non-Native Species in the Antarctic workshop, was now the chair of the Committee for Environmental Protection, and he started his two-year term with an ambitious agenda that was epitomized in the five-year work plan that he brought to the CEP. He had been elected, uncontested, at the 2006 CEP meetings. Henry also offered more direct political commentary from the head table of the CEP than his predecessor had: he voiced more support for the ASOC, the environmentalist NGO delegation I was a part of, and he encouraged wide-ranging discussions not only on the procedural issues of the CEP but in relation to the interpretation of the Madrid Protocol in a more general and philosophical sense. Henry had surveyed the CEP members over the year and from this had determined that biosecurity has come to be seen as a key concern for the committee. His five-year work plan suggested that biosecurity be made its own separate agenda item and that the CEP plan to devote an entire morning or afternoon session on this topic at the next meeting. The CEP supported this.

CEP member states presented information papers that addressed various dimensions of Antarctic biosecurity: "Non-native Species: Pathways and Vectors between New Zealand and Scott Base, Antarctica" (New Zealand), "Hull Fouling as a Source of Marine Invasion in the Antarctic" (SCAR), "The Global Invasive Species Database" (New Zealand), "IPY Aliens in Antarctica" (Australia, SCAR), and "Prevention and Management of Harmful Non-native Species in the Antarctic and the Sub-Antarctic" (IUCN). When one of these papers came up on the CEP agenda, Henry would refer to the non-native species workshop of the previous year, reminding the group that it had identified biosecurity as one of the most critical emerging Antarctic environmental issues and mentioning the idea of giving the topic its own agenda item and dedicated discussion time at the next CEP. While this sounds repetitive, his commentary was reproduced and consolidated into the discursive patterns of the CEP and the wider ATCM, where consensus is attained not in a split second but through (typically) gentle suggestions that offer numerous opportunities for (typically) polite disagreement. Henry's leadership on this topic in his new role as CEP chair encouraged further momentum. The stream of biosecurity papers presented at the 2007 ATCM demonstrated that the outcomes from the Non-Native Species in the Antarctic workshop brought forward a conversation that is likely to lead to increasingly developed international policy.

Unlike the anxiety that sometimes arises during the implementation of biosecurity practices at the Antarctic borders that I described in the previous chapter, international policy making on this issue is not characterized by anxiety at all. The cultivated relationships of the participants in the Antarctic Treaty System and the expectation that states work in the best interest of the Antarctic drive the regime. The scientific work that informs these international policy documents show up again as short sections of the more general ATCM final report that must address scientific research, the political needs of states, and adoption by consensus.

Relying on the Form and Matter of Scientific Practice

Consensus-based policy decisions about Antarctic environmental management are produced in texts in particular settings (see Fish 1980; Jasanoff 2004; Jasanoff and Martello 2004; Latour 2004; Riles 2000, 2006). In the production of these texts, personalities, luck, surprise funding opportunities, and information presented in a way that captures attention are as important for the process of the movement of scientific knowledge through the policy system as rigor and peer review are. The scientists and policy makers who attempt to push research-based policy use writing and negotiating methods that privilege particular kinds of knowledge and particular technocratic skills.

The production of documents is one of the key types of work people do in the Antarctic Treaty System. Indeed, sometimes some text on a paper is the only meaningful work that is achieved in this slow-moving, diplomatic, and consensus-based arena. At a minimum, a group effort to negotiate the text of a report is a way to bring epistemic concerns into formal policy making through explicit and implicit maneuvers.

Though the Antarctic Treaty Consultative Meeting is not explicitly economic, the institution nonetheless exhibits neoliberal flourishes, in which the economic content is minimized in discourse and practice but the governance priorities remain, including an interest in future economic possibilities. One example is the primacy of the ATCM as a venue for information sharing among parties as a move to prevent censorship and in compliance with the idea of transparency as a "bureaucratic virtue" in neoliberal regimes (Hetherington 2011, 3). As making data available to the public has increasingly become the norm in national science programs, especially due to the proliferation of online data storage, data about research and logistics have become information to share, denoting industrious virtue in technocratic activities.

Modern policy making is driven by scientific data in all arenas. Political polls, statistical analyses of policy outcomes, and measures of which problems exist all rely on the perspective of Western science to help policy makers make decisions. Foucault's (1998) notion of biopower discusses the scientific ways of knowing populations as a way of asserting power over nature and of making the exertion of power seem natural. The techniques of science, as Greenhalgh (2008) argues, are the very means by which "regimes reason" (xv). Scientific input in policy processes is well documented, as are the roles political bodies and political decisions play in scientific research. Governance cannot be effective without a fair amount of data that supports decisions and provides credibility and a sort of objective legitimacy.

In contemporary science-based policy making, science is not merely an ingredient but a discursive and technical means for elaborating policy decisions. While conventional wisdom suggests that bureaucracy is where good science advice goes to die, this chapter suggests the converse: scientists and their research drive technocractic practices, enable or hinder the work of policy-making bodies, and create institutional space where science is referred to (see also Hull 2012). This is not to say that the translation between science and policy is impartial and perfect, but that scientific practices are transformed through technocratic institutions and bureaucratic work.

Reference to and Reverence for Science: Working Science

Personal, real-time negotiations fall away somewhat as the project of biosecurity travels into international policy. In its place, formal, state-level concerns flourish alongside the usually slow-moving process of achieving complete consensus in Antarctic Treaty consultative meetings. In the case of biosecurity, the issue was helped along by the Non-Native Species in the

Antarctic workshop in Christchurch, New Zealand, where various representatives mobilized their expertise and hashed out their commitments and suggestions. The results from the workshop were translated into a working paper at the 2006 ATCM and encouraged a proliferation of responses at the 2007 ATCM. Scientific advice became sound bites and lists that included proposals and calls for further study. This process illustrates how in the formal moments of policy making, experts become *reliant* upon science.

The Antarctic Treaty consultative meeting and its ancillary body, the Committee for Environmental Protection, are designed to address nationalist and environmentalist concerns through the annual implementation of the policy process. The process is meant to consolidate Antarctic problems into agreed-upon, uniform solutions. But as the case of biosecurity illuminates, nationalist and environmentalist concerns related to a specific issue reveal other issues.

Lessons learned from the work on biosecurity help Antarctic managers, policy makers, and scientists figure out ways to make knowledge-based policy. While sometimes Antarctic science looks like any Western science, Antarctic scientists are called upon to be representatives of the place and to embody their research in ways that can allow their research to inform policy. Similarly, while Antarctic policy-making meetings are recognizable in relation to many other international fora, the policies are the result of attention to the specific, hard-to-get-to Antarctic environments and focused conversations with and among the scientists who are the most intensive human presence on the continent. These conversations convey the needs of Antarctic scientists and policy makers vis-à-vis the place upon which these people focus their attention and which they are expected, to some degree, to protect.

International Antarctic policy is tenuously hinged on complete consensus, and this fragile mode of agreement requires specific, usually gentle and polite, modes of negotiation. While the Antarctic Treaty is often upheld as a model of international cooperation, the work involved in facilitating this is couched in idealism, pragmatism, and something similar to a leap of faith. The work accomplished under consensus is slow and is driven down to the lowest common denominator of agreement. It often results in the production of scoping documents and action plans that are edited for years in the hope of shaping the actions of national Antarctic programs.

In order to make consensus-based decisions on how to manage the Antarctic environment, managers produce texts. This chapter looked at various texts that were shaped during a season of discussion about biosecurity policy—a typed list of discussion questions at the New Zealand nonnative species workshop, the PowerPoint slides that workshop participants filled in, the working paper on biosecurity presented at the Edinburgh CEP meeting, the ATCM final report, and the hypothetical biosecurity measures of the Antarctic Treaty System that are expected to eventuate.

Because these texts are meant to be based on sound Antarctic science, policy makers select scientists to represent the perspective of researchers, which is, of course, neither monolithic nor static. This selection of stakeholders is not random; selection of policy participants must be seen as part of the production of policy decisions. Who is chosen and why? How does a scientist succeed or fail as a representative spokesperson? Who is excluded and why?

Decisions such as who to select as participants to contribute to an early-stage policy workshop are part of the management of the environment. We cannot separate the cultural, human dramas from the scientific or wilderness ones. Small, almost silent decisions at the grassroots level filter up into the highest level of policy decision making. Conflict is reduced, managed, and eliminated at the outset of the policy process: only particular spokespeople and their specific visions of the Antarctic environment filter through to the binding policy texts that are eventually produced. In essence, the Antarctic people make the Antarctic into an epistemic technocracy as they do the double work of writing science-based policy and negotiating it through the idealistic and ponderous web of international Antarctic governance.

6

TECTONIC TIME AND SACRED GEOGRAPHIES
IN THE LARSEMANN HILLS

At the 2006 Antarctic Treaty Consultative Meeting, I spent hours listening to procedural issues that offered few surprises. As the participants worked through a packed agenda, a delegate from the Republic of India announced that India was planning to build a new research base, its third, in the middle of the proposed Larsemann Hills Antarctic Specially Managed Area.[1] An Antarctic specially managed area, or ASMA, is a management category that implements specific policies to protect the environment in heavily visited Antarctic places. Australia, Russia, Romania, and China had been working on the ASMA draft management plan since 1999, and many of the delegates felt that the plan was ready to be approved.[2] The ASMA management team already had bases in the Larsemann Hills and was hoping that the establishment of an Antarctic specially managed area in that area would prevent future development and put stricter environmental regulations into place. While the existing bases are concentrated in one area, the proposed Indian base would be situated in the center of an undeveloped area, miles away from the other bases. India's announcement ground the ASMA management plans to a halt.

The Indian delegate spoke for several minutes, outlining the prospective scientific activities that would take place at the new Indian research station. He explained why it was so important that India's Antarctic program locate this base in the center of a proposed Antarctic specially managed area, in a mapped spot designed to prevent this type of development from taking place. This area, he said, was important to the nation of India because this was the point where the Indian subcontinent and Antarctica were connected during the time of Gondwana, 125 million years ago.[3] In addition, Indian geologists had suggested that the promontory they had selected in the Larsemann Hills was where the Indian holy river Godavari had flowed when the landmasses were connected. Because of this, he explained, India had a "scientific need" to build a research station at this site to study Indian-Antarctic geologic relationships.

Some delegates were furious about this announcement. During the Indian delegate's presentation, people began leaning over, speaking to one another, and tilting placards with their states' names on them, indicating their desire to speak. A long session ensued as each delegation was given a turn to voice its opinion. A few supported the Indian program's plans, but most were politely critical. Delegates mentioned the environmental impact of having so many bases in one area. Many people discussed the wrench this threw in the ASMA management plans and the years of hard work that other Antarctic Treaty members had put into this plan. In casual conversations after the formal meeting, some people suggested that Indian scientists were merely proposing to conduct research that had already completed by other scientists. Others opposed to the new Indian station suggested to me that "India doesn't understand the process" of Antarctic policy making. Because the Indian delegates had followed the formal Antarctic Treaty rules of procedure, these critics seemed to be suggesting that another, informal process had not been followed. This process that India is purported to not understand, I argue, is what happens in the gap between national agendas and global governance, where people in one national Antarctic program must translate their plans into internationally acceptable policies.

This translation often takes form in scientific language that is performed strategically in the formal meetings, in sideline negotiations, with gossip, and in the press. Although some suggested that Indian policy makers did not understand the process, I contend that Indian Antarctic delegates did understand it and finessed it in ways that pushed against the norms many

state delegates followed. The Indian delegate's deployment of sacred geographies and tectonic time is a tactic that positions him and the Indian
Antarctic Program as both expert insiders and outsiders to the mainstream
of the Antarctic Treaty meetings, scientifically and geopolitically.

The Production of Policy

The Antarctic Treaty System as a system of governance contains multiple layers and both explicit and implicit processes for making policy within
these layers. Systems of governance in action, however, are not simply organizations in which authority moves in a singular, simple way. Riles (2000)
refers to bureaucratic systems as "networks," "systems that create themselves" (173). People in these systems not only create workable networks but
sometimes become problematically entangled in them. Decisions do not always slip into and out of policy documents with ease: the creative work of
the bureaucracy relies on agreement among the people participating in the
process. This accord is not always easily achieved. Using alternative processes to work through political intransigence via is a well-documented process (for example, see Chayes 1996), as are the dynamics of postcolonialism
and the severe imbalances of power in global governance (Anghie 2005;
Kennedy 2008). This case reveals how scientific discourse is both opened up
and disciplined through policy processes in a global venue.

During the mundane procedures of governmental and nongovernmental organizations, people produce knowledge (particularly procedural
knowledge that interprets policies), alongside the documents and relationships that are forged or tattered in meetings. Negotiating policies involves
wrangling over text and numbers and interpreting them into action, re-
writing as needed. A focus on policy writing draws attention to the modes
of power that are mobilized in the process of writing policy documents.
These modes involve strategies for managing difference and agreement
in the text (see Riles 2006). Foucault (1977) famously suggested that modern institutions situate people in "networks of writing," where inscriptions
emerge from sets of disciplinary practices. Brenneis (1999) and Strathern
(2000) have built upon this idea, showing that bureaucracies are rich sites
for producing new knowledge and enveloping this knowledge into the everyday and repetitive practices of bureaucratic work.

Going through the process of moving policy through the Antarctic Treaty System in ways that are socially acceptable to dominant parties within the system becomes as—or more—important than the content of the policy.[4] This chapter argues that the representatives of the Indian Antarctic Program were working with an understanding of the dynamics of postcolonialism and nationalism within the Antarctic Treaty System. The Indian delegation circumvented the ASMA process and worked on the basis of their understandings of how power is used in the Antarctic Treaty consultative meetings.

Although proceeding in accordance with other delegations' wishes would have made India look like a good member of the Antarctic Treaty System, it would have undermined the Indian state's ability to shape its Antarctic research and logistics strategies. This is another consideration: the Indian Antarctic Program needs to convince Indians of its value, especially for the purpose of garnering funding appropriations. The states that would most likely have been expected to consult with and advise the Indian delegation about site selection and logistics for their new base are the most well-established, wealthy, and powerful national Antarctic programs. While their advice would have been couched in collegiality, information sharing, or environmental concerns, it also would have been neocolonialist. Indian policy makers understood this. The Indian Antarctic Program subverted this by refusing to participate in the process that opens up opportunities to be bullied and pressured. Instead, managers from the Indian Antarctic Program did not make their new base plans public until their national government had approved and committed to them. Akhil Gupta (1992) describes how the specific kinds of transnationalism that emerge among developing nations are different from the dominant internationalisms wealthier nations promote. A counterhegemonic policy approach—refusing to full comply with informal, moral, and bureaucratic norms—can be successfully cultivated by less powerful states. In this case, Indian spokespeople appropriated scientific discourse and reconfigured notions of time and space to legitimize its claim to Antarctic territory.

From Formal Politics to Informal Process

A range of informal and formal policy-making processes makes international management possible. The international body of Antarctic policy makers relies on consensus about appropriate environmental management. Much of the political legitimacy and authority of each state at the Antarctic Treaty consultative meetings is related to its scientific stature, broadly construed. Territorial claims and colonial arrangements past and present still matter heavily in the Antarctic Treaty System. However, the delegations bind these problems into the language and practice of science.

After India's announcement, the tone and direction of the discussion about the proposed Larsemann Hills ASMA shifted. One of the most intriguing details of Antarctic specially managed areas is that they force a contingent of several nationalities to turn a seemingly blank Antarctic space into a decidedly international, managed one. ASMA management teams are comprised of nations with bases in or near the ASMA. As soon as the ASMA management plans begin to be written, the involved parties negotiate procedures and regulations for how to live in, develop, study, visit, and categorize a piece of Antarctic land. While all of the Antarctic is international, it is also large, spread out, and relatively unpeopled. ASMAs concentrate the roles and ideas of nations onto a relatively small international space. New and tiny national borders seem to become drawn around national bases, and the national representatives must share resources such as runways and agree on environmental and other policies.

The consternation of several Antarctic national delegations about the Indian proposal indicates that the gap between the strategy India used to convey its national interest in building a new base and the rhetoric and practice of Antarctic environmental management was not an easy one to bridge. Although the Indian delegate's Gondwanan rationale was framed in scientific terms, it was not accepted as such by other national delegates.

Tectonic and Sacred Geographies

As the delegates from the Republic of India mobilized arguments for building their new base in the Larsemann Hills, they reimagined the kinds of expert evidence that Antarctic decisions might be based upon. The people working on the Larsemann Hills ASMA management plan and those working on the plan for the new base had to find ways to work with two clashing time scales: that of contemporary international geopolitics and that of the Gondwanan megacontinent.

One of the most compelling justifications for the location of the new Indian base in the Larsemann Hills is that it is the site where India and Antarctica were joined when they constituted part of Gondwana. Contemporary claims and longings for ancient places, though idiosyncratic, have been presented before. In *The Lost Land of Lemuria* (2004), Sumathi Ramaswamy traces the histories of a supposedly submerged continent and land bridge in the Indian Ocean. The "lost continent" of Lemuria has been theorized over time by European scientists, American New Age adherents, and by the Tamil in India and Sri Lanka, who claim that Lemuria is the (now submerged) Tamil homeland. Although Lemuria is a different "lost" landmass from Gondwana or Pangea, particularly because now most scientists do not think that Lemuria ever existed, Ramaswamy calls Lemuria a "paleo place world" like Gondwana and looks at "fabulous" and "catastrophic" histories of Lemuria and the overtones of loss such histories carry in contemporary narratives. The Tamil people (who live in southeastern India and northeastern Sri Lanka) vociferously claim that Lemuria is their vanished homeland, that it was a wide-spanning continent where the first humans came from, and that these humans spoke Tamil. Ramaswamy (2004) explains that Lemurian place making is rooted not in ancient myth and supposition but in early Western scientific disciplines. She writes that "in an age mesmerized by the global revelations of the paleo-sciences, Lemuria was a creation of those very sciences" (5). As fantastic as Lemurian narratives appear to be, they are grounded in scientific exploration and hypotheses about the distribution of fauna in Africa, India, and Madagascar. A British biogeographer named Philip Lutley Sclater (2004) first proposed and named Lemuria in 1858. The scientific hypothesis of a land bridge that became a highway for the transmission of fauna was in play, though in a marginal way, from the mid-1800s until the mid-1900s, when more definitive evidence for plate tectonics became convincing.

The notion of the lost continent of Lemuria eventually became discredited in geological studies over the course of the twentieth century. Ramaswamy, though, maintains that the work related to the lost continent—particularly by the Tamil—is a commentary on the scientific work that discovers lost continents. She writes, "The labors of loss around the paleo place-world of Lemuria are not vestiges of the archaic (although in many cases, they may appear so, and strategically), but are the outcomes of and responses to the various projects of scientific and colonial modernity as these came to be conducted across the inhabited world" (2004, 9). Lemuria as a paleo world place inhabits a particular tectonic geography that interacts with the lived geographies of contemporary people, a way of using history to speak to scientists and the state. It just happens that it did not exist, according to most scientists. Lemuria was a means by which the Tamil could position themselves alongside and against Western science, using tectonic time.

Gondwana was another large and ancient landmass, first named by Austrian Eduard Suess in *Das Antlitz der Erde* (*The Face of the Earth*; 1904). Unlike scientists' views of Lemuria, the scientific community came to agree that Gondwana did exist, although the acceptance of the theory of continental drift has its own complex history (Oreskes 1999). The naming of this landmass is significant, as are the exploratory activities that followed its naming (or discovery, if one can label a no-longer-existing place as discovered). Paul Carter (1987) suggests a key difference between discovery and exploring: discovery entails finding, mapping, and naming a new place, while exploration involves coming to know a specific place well (25). The science and politics of the Gondwana paleo place world continue to be bound up in exploratory work, as people try to come to know the place and explain it scientifically.

In the case of the Larsemann Hills, while Indian exploratory work is indeed scientific, it is also wrapped up in the work of nationalism, territoriality, and religion. This work is similar to the claims about Lemuria the Tamil people use to negotiate their standing in a postcolonial, scientific world. This is not exclusive to Indian people and it is not a mystified response to modernity: this negotiation of science, nationalism, and postcolonialism is embedded in all national Antarctic projects. It is not just Indian managers who deploy alternative strategies to situate themselves in the Antarctic: all Antarctic national representatives use science strategically to advance their national interests. However, the Indian delegate's deployment of arguments about religion and deep geologic time was strategic in two ways. These arguments contrasted with the dominant vernaculars of Antarctic projects, thus creating an alternative pathway for policy within the Antarctic Treaty System. And the members of the Indian Antarctic Program knew that arguments about sacred geographies and tectonic time would be convincing for many Indian citizens and would create a sense of national purpose for their proposed plans. The types of arguments that are typically used in the international arena would not necessarily have been as compelling in gaining Indian national support.

Strategic Science and the Reappropriation of the Policy Process

In India, nationalist and media representations of the proposed new Indian base highlighted the spiritual and geo-nationalistic claims about the Gond-wanan connections between India and Antarctica at the Larsemann Hills. Various members of the press treated the notion that the Godavari River had once flowed through the Larsemann Hills as absurdly silly and utterly serious. Today, the Godavari River flows through the states of Maharash-tra and Andhra Pradesh and is considered a Hindu sacred site. There are several pilgrimage places along the banks of the Godavari where impor-tant Hindu and Indian figures have bathed over thousands of years.

In the earliest days of planning the new Indian Antarctic base, represen-tatives of the Indian Antarctic Program mentioned this geologic and sacred link between India and Antarctica often, particularly in the national press. However, the point that the Godavari River in Antarctica was sacred did not go over well in the international press or in the Antarctic Treaty con-sultative meetings (Jayaraman 2007; Ramachandran 2007; Sharma 2006, 2007; Mallikarjun 2003; Kamat 2006a, 2006b). Like the fact that India and Antarctica used to be connected millions of years ago, this sacred geogra-phy had to be retooled as a piece of geologic evidence rather than a claim to nationalistic, spiritual Antarctic places for the Indian people. Situating this part of the Larsemann Hills as a sacred Indian geography implied a longer timescale of legitimacy and a deep, inalienable relationship between Indians and this piece of Antarctica. Journalists did much of this situating through communication about press releases, Indian research reports, and interviews with the controversy's key actors.

Both the Indian press and the international press were interested in the dispute within the Antarctic Treaty System the Indian claim generated and described the case differently. National press about the new Indian base focused on efforts to gain and sustain public support and interest. In *The Hindu* on September 2, 2003, a headline read "India Plans to Set Up Permanent Base Station in Antarctica," and the article that followed de-scribed an upcoming trip to search for a location for this new base. The journalist Y. Mallikarjun quoted Harsh K. Gupta, the secretary for the Indian Department of Ocean Development, as saying that this new station would be "huge" in comparison to the previous Indian Antarctic bases. On May 26, 2006, reporter Prakash Kamat (2006b) wrote an article just before the Antarctic Treaty Consultative Meeting in Edinburgh about a national seminar on Antarctic science. Kamat reported on the scientific potential for India in Antarctica, which included medical bioprospecting, fisheries research, climate change research, and possibly a solution to tsunamis. After the fractious reaction at the ATCM, though, Indian journalism focused less on information about geography, science, and the nationalistic value of research in Larsemann Hills and more on India's controversial positioning. A news update in *The Hindu* led with this sentence: "India is set to defy international opinion and set up a base in a planned Antarctic protected area to research the prehistoric origins of a Hindu holy river" (Kamat 2006a). In this series of articles, India's claim was as much about sacred geographies as it was about internationalism and world positioning. After making these claims in the press and at the ATCM, the Indian Antarctic Program and other states interested in this problem had to situate these alternative geographies in the specific, small-scale, and managerial geographies of Antarctic specially managed areas.

As Indian journalists and policy makers conveyed the different rationales of sacred time, geologic time, and scientific research to their audiences, they engaged with the larger process that holds sway in international Antarctic activities: if states pay close attention to the procedural, bureaucratic activities involved in the approval process, they can almost always do what they wish in the Antarctic with little, if any, intervention.

In an interview in Hobart, Australia, I asked the Indian delegate Rasik Ravindra to explain the process by which India decided to build a new station and why India had decided upon this location in the Larsemann Hills. We sat in a nearly empty corridor at the hotel that was hosting the Scientific Committee on Antarctic Research conference. Many people who had attended the ATCM in Edinburgh a few weeks earlier were here, jetlagged and trying to shift gears from environmental policy to the details and logistics of science programs. He explained that the Indian Antarctic Program wanted to expand its research to include more than one site in order to make the research more robust and environmentally diverse. They also wanted to find a site more appropriate than Maitri for

conducting oceanographic studies. To find the site for the new base, Ravindra said, the Indian Antarctic Program cruised the Antarctic coastline
for 2,000 kilometers to consider three different places: the Antarctic Peninsula, the Bellingshausen Sea area, and the Amery Ice Shelf. They decided
against the Antarctic Peninsula because it already has so many bases and
they did not want to contribute to the human impacts there. The Bellingshausen Sea location was not feasible for a year-round station because of
the sea ice conditions there. After wide-ranging travel in the Amery Ice
Shelf area, the team decided that the Larsemann Hills site was the best
location. They did not want to be too close to the Australian base because
the Australian researchers there had already done so much work in the
area; the Indian Antarctic Program wanted to pursue new research directions. They rejected a few locations because of their proximity to nesting
bird sites: they wanted to avoid impacts on these habitats. After Ravindra
detailed the logistical considerations, he described India's scientific interest
in the area, which included but was not limited to the Gondwanan connection between India and the Larsemann Hills (interview, July 13, 2006).

Ravindra's description of the process by which the Indian Antarctic
program considered other locations for bases conflicts with arguments
that India did not adequately consider other options. His narrative demonstrates a consideration of other options, though the rigor of these considerations is impossible to gauge. Ravindra also claimed that the decision
about where to locate the base was motivated, in part, by a desire to avoid
repetition of research that had already been conducted by others.

Furthermore, Ravindra argued that the sacred geography implied in
the geohistoric link between India and Antarctica was a main *scientific*
reason for constructing a base in the Larsemann Hills. Because the two
landmasses were previously connected, Ravindra explained, much can be
determined about the geology of both regions. The entire geologic story
of this piece of Gondwana, Ravindra said, cannot be told only from India
or only from the Larsemann Hills. It is the interactions of the landmasses,
told through geologic observations, that can explain the processes that occurred during the Gondwanan time period and the time when present-day
India split off from present-day Antarctica.

In this conversation, Ravindra asserted that the Gondwanan connection between India and Antarctica is scientifically important. In doing so,
he set aside his claims that the connection between Antarctica and India is

nationally compelling. This is an illustration of how scientific arguments bring perceptions of legitimacy. India's arguments could be successful in this policy context only when they could be cleared of sacred, nationalistic, or territorial overtones and reframed scientifically. In the 2007 ATCM in Delhi, the Indian delegation presented this reframed justification for their new base and it was approved.

Near the end of our interview, I asked Ravindra to comment on the consternation and controversy that emerged at the Antarctic Treaty Consultative Meeting when he announced the location of India's new base. He responded in a diplomatic manner:

> First of all, I don't think there should have been any controversy, and in fact from our point of view there is no controversy. Antarctica is a continent of peace, tranquility, and we are a signatory to the Madrid Protocol. We are the people who want to maintain the scientific significance of Antarctica, the pristine environment of Antarctica. From that point of view, we don't find that there is any controversy. The only point which has been raised in the last ATC meeting was about the ASMA: that is, the Antarctic Specially Management Area plan by Australia and other countries. And [we] think that will also be sorted out because we told them that we are not against the ASMA, we joined the ASMA consortium. We will begin to manage and protect this area. Staying away, possibly one cannot do much, but if you are part of the management activity, possibly you can do much more.

Ravindra did not mention the accusations that have been made about the Indian Antarctic Program or that his program had decided to build a new base after an ASMA had been planned in that area for the better part of a decade. Instead, he emphasized the environmental and scientific goals the Indian Antarctic Program shared with other states.

Perhaps most telling of India's understanding of the process in the matter of the Larsemann Hills ASMA was Ravindra's statement w when I asked him what he thought the outcome of the planned Indian base will be. Easily and matter-of-factly, he said, "Of course the base will be built." Although Ravindra has retooled the arguments for the new base so they are about logistics and science, he indicated here that national interests will be upheld regardless of whether they garner international approval.

Ravindra's statement that "of course the base will be built" reflects the reality of the situation. Simply put, despite the international treaties signed into law by the signatory states to the Antarctic Treaty, these states and nonmember states can do what they like in Antarctica (see Bastmeijer and Roura 2008). It is the domain of no one. The Antarctic Treaty is largely based on good faith, as it is politically costly to enforce treaty practices when states make small infractions. Ideas made to seem somewhat silly in international forums—such as using the existence of an ancient Godavari River to justify a new national base—can be retooled to resonate with scientific research activities and a universal desire to understand the technical minutiae of plate tectonics. Indian policy makers are certainly not the only people to use this strategy in their contributions to Antarctic policy and in their descriptions of their nation's projects on the ice.

In India, as elsewhere, science is marketed among nationals as a mode of governance. Scientific "conspicuous technology" helps legitimize scientific activities for the Indian middle classes, as Nandy (1988) has shown. During the development of the Indian nuclear program and its subsequent test of a warhead, scientific work was done as a nationalistic project in order to lay claim to political positions in international relations. India demonstrated its military capabilities and promoted arguments about science that was being conducted in tropical places away from colonial metropoles. India's national bureaucracies repackaged science, wrapping it in the ideologies of postcolonial nationalism, as Itty Abraham (1998) explains. Nationalist justifications for scientific research helped create public support. For the international public, though, different justifications may be needed.

Indian spokespeople provided rationales that were keyed to the spaces where they presented them: in the formal Antarctic Treaty Consultative Meeting space, which required less attention to scientific detail; in the sideline negotiations of the meetings, where people probed scientific argumentation much more closely; in nationalistic spaces in the media, where the argument was framed as postcolonial conflict. Indians often presented what some might see as more "credible" (i.e., scientific) evidence in backstage settings, while they used less strong scientific arguments in the more official settings because they could not be contested outright in those arenas. The official meetings, it must be noted, exist to enact politically adequate and tolerable decisions among all of the delegates. While delegates often issue calls for a vague "science," the work of consensus is a tenuous alchemy that uses a full range of sociocultural considerations.

The Indian Antarctic Program worked on generating national will before introducing its intentions to the international Antarctic community. Its plans, which were primarily communicated through the national press, described the geological and religious connections between the Indian subcontinent and the location Indian scientists had chosen in Antarctica for the new base. A constituency was raised as Indian nationals were convinced of the legitimacy of this project (see Jasanoff 2005). When India framed its rationale for the location of the new base in the Larsemann Hills, it used tectonic time to put into play a much longer time scale of relationship between a nation and the Antarctic. Paleogeological relationships between India and the Larsemann Hills infer a long-term notion of territoriality that contrasts with the more conventional time spans many Antarctic states use to create nationalistic political legitimacy and authority. While this deft deployment of regional mythology, popular culture, and notions of what science can mean for a national society succeeded in generating the requisite public support within the Republic of India, these claims failed to tie into international Antarctic rhetoric.

Meetings on the Margins and Other Tools of Consensus

At the 2006 ATCM, some delegates used technical means to present their opposition to the new Indian base. This strategy is more conventional (and thus more acceptable in this international arena) than the alternative legitimizing strategies India used. Organizations that build policy by consensus rely on various techniques to grapple with discord and make issues acceptable enough for the body to function. The Committee for Environmental Protection had to work out a response to the Republic of India and to the states involved in the initial Larsemann Hills ASMA management plan: Australia, Russia, Romania, and China.[5] By the end of the meetings, a few paragraphs of official report text had been constructed (for a somewhat analogous case study, see Riles 2000). The production of this text was the result of editing during the CEP meetings, negotiations that took place in meetings on the margins, and an orchestrated intervention.[6] Activities on the sidelines of the official, ritualized, and procedural steps of the Committee for Environmental Protection and the Antarctic Treaty Consultative

Meeting generated the eventual negotiated text. The formal meetings put into documents the product of negotiations during tea breaks, casual conversations, intersessional work, and other informal and ad hoc moments (Riles 2000; Scott 1998).

After India announced its plans to build a new base in the middle of a proposed protected area, the chair of the Committee for Environmental Protection instructed delegates from the ASMA management team and India to meet during tea, an example of an off-the-record meeting on the margins during which people from multiple delegations try to reach consensus about an issue and compose appropriate report text. Writing policy and report text is separate from the work of presenting, questioning, and approving this text, so the majority of this writing must occur before a meeting or on the sidelines of a meeting.

At the Larsemann Hills meeting on the margins, six delegates met in a cluster at the rear of the CEP meeting room, near the translators' booths. One delegate I spoke with explained that the ASMA management group's goal was to get their management plan through to the ATCM plenary at these meetings. During this brief but intense meeting, an ASMA management team delegate suggested that India write a revision for the ASMA management plan that stated that India might add an operational zone later. An Indian delegate replied that India wanted to wait and resubmit the ASMA management plan the following year. Another ASMA management team delegate said that they had been working on this plan since 1996 and would really like to see it go through that year.[7] The parties decided to meet later, since they were not able to agree during the fifteen-minute tea break.

The next day at the CEP meetings, the Indian delegation presented its working paper on its new base in the Larsemann Hills. Before Indian delegates began their presentation, the CEP chair asked for an update on the ASMA management plan from Australia, and an Australian delegate announced that they were no longer seeking endorsement for the plan, that they needed more discussion over the next year among all of the stakeholders, including India. The Indian delegation had decided to join the ASMA management team and had agreed to help rewrite the management plan, taking the new Indian base into account. The members of the newly configured AMSA management team had made plans to meet at another set of international Antarctic meetings in Hobart, Australia, in July.[8]

At the end of the meeting, the CEP finalized the committee's report before sending it to the larger Antarctic Treaty Consultative Meeting plenary. Paragraph by paragraph, the committee edited the information that would be sent to the plenary session. The process of adopting the report became quite interesting (this activity is notoriously tedious in general) when it came time to discuss the Larsemann Hills ASMA management plans section. One national delegation coordinated an intervention to publicly complain about the insertion of a new Indian base into the ASMA. This intervention was a surprise to the Indian representatives in a space where surprise is rare; it caused some embarrassment and required some explanations in the more public space of the general meeting assembly.

When the subject of the text about the new base came up again during report adoption, a speaker stated that his delegation just found out that India had already erected a building—a portable structure called a melon hut—on the proposed new base site without filing any environmental evaluation paperwork, as required by the Antarctic Treaty and the Madrid Protocol. After some tense discussion, the chair told the delegations involved to meet during lunch to work out the text.

After lunch, the drafting group offered new paragraphs for the report. The Indian delegation claimed that India did not have to do an initial environmental evaluation because its Antarctic national program was drafting a comprehensive environmental evaluation. The Indian delegation claimed to be obeying a different order of rules—a gesture of compliance that worked well enough that the other states could credibly accept it. There was nothing in the paragraphs about the melon hut, since nothing technically could be done about this situation and perhaps there was no will to embarrass the Indian delegation further. The paragraphs were adopted into the report. While the process was consistent with what usually takes place in CEP meeting rooms, the process on the ground in Antarctica was anything but consistent. Placing the melon hut without going through the plans with the CEP delayed international scrutiny and allowed the Indian national program to proceed without sanction, although its delegation had to endure some embarrassment in an international forum.

Although the meeting on the margins and the public intervention helped make the text of the CEP's report to the AMSA, there is no reference to such activities in the report. The result of the meeting on the margins was that the Indian delegation achieved its goal: the Larsemann

Hills ASMA management plan was set aside for another year and Indian activities in the area were implicitly given clearance, despite objections and a loss of favor among some of the powerful delegations in the Antarctic Treaty System. The Indians used the meeting on the margins to insert themselves and their plans into the policy process. In just fifteen minutes of off-the-record discussion, a seemingly inevitable international environmental management plan took a new direction. While the public intervention did not stop India's planned new base, it was an attempt to demonstrate to many of the Antarctic Treaty members that India did not understand the process. The tense situation surrounding India's plans for the Larsemann Hills underscores the ethnocentric assumption that only Westerners can be considered experts in conducting bureaucratic processes (Herzfeld 1992; Heyman 2004). The complaints that the Indians did not understand "the process" are most clearly weighted with the idea that somehow, Indian delegates are not succeeding at inserting themselves into bureaucratic culture, that they are marginalized (or that others would like to marginalize them) in international Antarctic decision making.[9] Instead, the Indian delegate deployed different strategies—including tactics that embrace marginality—to achieve the goals of his Antarctic program at these meetings.

Tectonic Time and Sacred Geographies in Policy

From the many people and states involved in the conflict over the new Indian base at the proposed Antarctic specially managed area in the Larsemann Hills, two dominant stories arise. In the first narrative, the Indian claims about the ancient connections of India and Antarctica during Gondwana, particularly the claims that the holy Godavari River flowed on both landmasses, bespeak an ancient sacred geography. In the second narrative (and this is the claim of many who oppose India's plans to build a new base in the proposed protected area), Indian policy makers do not understand the process, perhaps due to cultural or linguistic misunderstandings (even though Indian delegates speak English, an official Antarctic Treaty language).[10] But there are also hints of a lack of institutional knowledge, a lack of expertise regarding the conventions of the Antarctic Treaty System and the careful, polite negotiations that occur there.

While the press picked up on the Indian delegation's initial argument that they considered the geologic landscape of the Godavari River to be sacred and ancient, populist and nationalist talk was largely subsumed by the attempts of other parties to enlighten India about the Antarctic Treaty process and its apparent failure to comprehend it. In short, the Godavari River argument was marginalized in the Antarctic Treaty System: it was met with jokes and rolled eyes. India responded by turning that narrative into a story about a compelling scientific need for the country to locate a base and scientists there. The basis for the claim about this particular patch of the Larsemann Hills was processed in the international arena to be about science rather than about territory, although the persistent claims about the Gondwanan continent of course are both. The management process allows and even encourages idiosyncratic arguments because only national representation—not scientific consensus—is required. This community agrees that arguments must be couched in scientific terms instead of nationalist ones: there does not have be epistemic accord with the scientific ideas presented. When the formal and informal processes are followed, there is little room to negotiate on scientific matters.

India's science-based claims were only partially successful, because some suggested that its planned scientific activities for the area merely repeated the work that had already been conducted by others with research stations in the area. Some Antarctic states zeroed in on a critique of India's Antarctic policy-making process. The formal rules of procedure for the Antarctic Treaty meetings are different from the informal process for achieving the goals of the meetings. The formal process involves following the procedural rules with ritualized flourishes that remark upon international cooperation, good faith, and good manners. It is a stylized ritual that allows national Antarctic programs to proceed as they wish as long as they do not flagrantly undermine the idealistic goals of the Antarctic Treaty. Thus, science packaged correctly can be deployed as a rationale for nonscientific motivations such as national pride and the promotion of national interests.

The process that some claimed the Indian delegates did not understand involves submitting activities for consultation and advice from the other Antarctic Treaty consultative parties. If the Indian delegates had followed the conventional steps involved in building a base in Antarctica, their opponents argue, other states would have been satisfied because the Indian delegation would have appeared to have acted in good faith and as good members in a body that emphasizes international cooperation and information sharing.

The actions of the Indian Antarctic Program in this case also subverted the way that claiming takes place in Antarctic matters, even though (at least on paper) territorial claims are not to be exercised. Antarctic claims are made through a long-standing scientific national presence, state-of-the-art scientific programs, due diligence in environmental management, public outreach, and an international profile. Most Antarctic states have mastered many of these features, but few Antarctic states can demonstrate these features over the long term in Antarctica. Because people reportedly did not see the Antarctic until the 1700s, a long-term Antarctic human presence is usually considered to be less than 100 years and often is benchmarked by the International Geophysical Year of 1957–1958. The few states that have maintained a physical presence in the Antarctic since the International Geophysical Year also have authority in the Antarctic Treaty consultative meetings due to their longer experience in the region. This authority is moral, not official; their official status is equal to that of any other signatory state in the Antarctic Treaty System. However, their advice and their positions in relation to the activities of other states in the Antarctic shape the continent's future. States with less influence must rely on their ability to block consensus decisions at Antarctic Treaty consultative meetings and use other strategic tactics that destabilize the influence of long-term Antarctic states.

When the Indian delegates framed their rationale for the location of the new base in the Larsemann Hills, they used tectonic time to put into play a much longer legitimating time scale of relationship between their nation and the Antarctic. Paleogeological relationships between India and the Larsemann Hills infer a long-term cultivation of territoriality that contrasts with the more conventional definition of what constitutes long-term cultivations of expertise and human presence that many Antarctic states use to create nationalist political legitimacy and authority.

The Indian Antarctic Program demonstrated that national scientific rhetoric does not have to mimic prevalent international discourses. Discordant rhetoric is smoothed out in the gap between national agendas and international management documents, in what Antarctic bureaucrats call "the process"—consultation, the submission of paperwork detailing plans, and various modes of grappling with the concerns of other nations in an international forum. Thus, the international management strategies that emerge are the discursive and material products of the negotiations that take place during the policy process.

In transnational governance, producing policy is serious work. Novel ideas and arguments that counter standard procedures usually take one of two directions. Sometimes these outliers are ignored in efforts to produce the most generic, consensus-based policy document possible. Bracketing is a way of creating accord: issues that nations cannot reach agreement about are left off the table to be dealt with in an undefined "later." Alternatively, these ideas and situations can be translated into policy language that makes the unusual ideas fit. The story of the Indian delegation's announcement of its plans to erect a new base is one of these instances: the people involved chose to engage in the tough process necessary to somehow include, even partially, the Indian delegation's story in standard bureaucratic text.

7

Charismatic Data and Climate Change

"Pile-Up as Berg Hits Antarctica" —BBC News 2005
"Huge Iceberg Heading for Collision" —TV New Zealand 2005
"Get Ready for the Largest Demolition Derby on the
Planet" —Space Daily 2005
"The Iceberg Cometh" —Adam 2004
"Outta the Way, It's B15A!" —Eime 2005

As happens occasionally, an enormous Antarctic iceberg captured international media attention in 2004 and 2005. An iceberg, named B15, had calved off the Ross Ice Shelf into the McMurdo Sound in 2000. The size of B15, which was grandiosely labeled "the largest iceberg ever recorded," was compared to that of Jamaica and Luxembourg. In December 2003, when a large piece of B15 broke off, observers named this piece B15A. By 2005, B15A had become lodged in McMurdo Sound, blocking McMurdo Station and Scott Base from the open ocean. The berg blocked the approach route to McMurdo Station that icebreakers use, cancelling that summer's annual resupply trip. Both the United States and New Zealand Antarctic programs rely on the annual, U.S.-sponsored resupply trip to maintain their research programs.

Months wore on and the massive iceberg did not budge. Logistics managers in the United States and New Zealand began to arrange extra flights

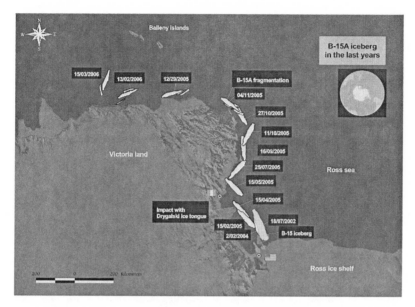

Figure 7. Movements of Icebergs B15 and B15A. Image by Luca Pietranera,
Wikimedia Commons file:B15a a4.jpg.

to resupply the bases by air. However, some scientists working in the area used the event as an opportunity for research. A penguin ecologist I interviewed called the event an "awesome experiment" and began studies detailing how Adélie penguin populations had changed: because the annual ice had not broken up, they had had longer travel distances to obtain food and colonies had relocated or diminished in size. He suggested that such events and their effects could become more common as the climate changes. An ocean ecologist studying the Ross Sea called the B15 blockage a "natural experiment" in how the local marine ecosystem responded to multiyear freezing.

In October 2005, B15A broke up into smaller pieces, making the annual Ross Sea breakup possible. Some claimed that waves from an Alaskan storm had provided the impetus that broke up B15A. In a piece on National Public Radio's *All Things Considered*, these small Alaskan waves were likened to the "butterfly effect," the term for the idea that a butterfly's flight can cause a storm on the other side of the world (Harris 2006). An Antarctic geologist in New Zealand suggested at a lecture I attended that the "Asian tsunami" of December 26, 2004, had managed to dislodge and break up B15A.[1]

The stories managers, scientists, and journalists told about B15 spoke to the concerns about climate change in Antarctica and the rest of the world, even though glaciologists considered the calving of B15 to be "natural wasting" (that is, not related to accelerated global warming). Managers worried about supplying their programs so they could continue running successfully. Scientists studied how the local environment and animals responded to local change as potential models of adaptation to climate change. Journalists invoked the potentially catastrophic imagery of a melting Antarctic continent and, by linking B15's eventual breakup to other world events, emphasized the interconnectedness of the world's climate system. The charismatic event of B15's calving off translated into stories and facts about global climate change.

Data never speak for themselves. They require spokespeople who not only report them but also bolster, finesse, and sculpt them to create sense for an audience. Through the presentations of experts, data can take on charisma, beauty, and amplified meaning. Charisma arises in relation to both scientific findings and the textual practices scientists use to explain their work. The way scientists, audiences, and policy makers handle data helps produce the political implications of global climate change and the strategies for mitigating it.

Facts are unstable. The notion that facts are socially constructed and require human intervention to make them legible and meaningful has a long history in science studies (for example, see Pinch and Bijker's foundational 1984 piece). Entire institutions work as "boundary organizations" to produce the apparent stability of facts (Guston 1999). The notion that facts come from data troubles traditional scientific narratives. In order to know where facts begin, one must consider the data.

Data do not exist in a natural state, though they often mark a scientific starting point. Lisa Gitelman (2013) explains that "data need to be imagined *as* data to exist and function as such, and the imagination of data entails an interpretive base" (4). Data form the origins of many scientific narratives of exciting discovery along with the careful and often tedious work of making data come to make sense in scientific practice. From data's emergence via technoscientific methods to its explicit interpretations in public presentations, people characterize nature through its quantitative, measureable forms.

How spokespeople handle data can imbue the information with particular, powerful qualities. To be a scientist-spokesperson is fraught with risk that includes the dangers to one's career and the more insidious threats to one's reputation. The idea that data need to be spoken for challenges the notion that truth has been read from nature without a mediator. As Choy (2011) notes, "Today we assume that knowledge is best when its spokesperson is no one in particular" (83–84). In other words, the ideal communicator of scientific knowledge is a particular kind of neutral, cautious, expert, what Haraway (1997) refers to as a "modest witness" (see also Shapin and Schaffer 1989). Haraway notes that scientists often position themselves as

> the legitimate and authorized ventriloquist[s] for the object world, adding nothing from [their] mere opinions, from [their] biasing embodiment. . . . [Their] narratives have a magical power—they lose all trace of their history as stories, as products of partisan projects, as contestable representations, or as constructed documents in their potent capacity to define the facts. (Haraway 1997, 24)

Haraway counters this with her reading of an appropriate scientific modest witness: "Valid witness depends not only on modesty but also on nurturing and acknowledging alliances with a lively array of others, who are like and unlike, human or not, inside and outside what have been the defended boundaries of hegemonic selves and powerful places" (269). The goal of this chapter is not to determine the source of the charisma of specific pieces of data but rather to suggest that charismatic data are co-constituted from the production of facts, natural events as recorded by scientists, and the ways spokespeople present these data to non-expert audiences.

It is not only the data that are transformed and gain power as they are presented by expert spokespeople. Through the discursive and academic moves scientists make in their presentation of themselves and their research, they can become spokespeople. This is a gradual process that scientists critically narrate, worry about, or strive for. The audiences at such presentations matter too: having an audience frames spokespeople as experts and fulfills some of the criteria for proper performance of public knowledge (Jasanoff 2004, 2005; Latour 1983). The invocation of charismatic data can provide a powerful bridge between the neutral presentation of scientific data and political advocacy. Peter Galison (1997) uses the term "golden event" to describe the production of an image of a physics phenomenon that is "the single picture of such clarity and distinctness that it commands acceptance" (22). A golden event is "an individual instance so complete, so well defined, so 'manifestly' free of distortion and background that no further data has to be invoked" (23). Visual presentations of charismatic climate change data have a similar power, but their power is amplified through collections of different sets of charismatic data. Presenters engage in specific cultural work with their audiences when they present visual representations of data as purified bits of information, free from distortion (Dumit 2004).

How is it possible that data can be charismatic? Scientific data can share properties with charismatic religious people, whom the faithful believe to have a direct link to spiritual and religious beings. To have charisma means to be favored, gifted, or imbued with an extraordinary power. In scientific practice, data about natural phenomena can become similar to what is known in conservationist circles as "charismatic megafauna"— cuddly, threatened animals that become symbols of the need to protect entire ecosystems. Dramatic examples of scientific data can "speak for" environmental problems more broadly. As with the B15 iceberg, the ways people assign stories and meanings to phenomena make them charismatic, as do the human allies who carry these data to public, policy, and scientific forums.

Weber (1947) defines charisma as "a certain quality of an individual personality by virtue of which he is set apart from ordinary men and treated as endowed with supernatural, superhuman, or at least specifically exceptional qualities" (329). However, Weber's body of work on charisma probes at his own definition. He notes that while charismatic people are exceptional, charismatics fill an ordinary need of cultural groups for regulating and organizing social life. Weber writes of "the routinization of charisma" whereby "in its pure form charismatic authority may be said to exist only in the process of originating. It cannot remain stable, but becomes either traditionalized or rationalized, or a combination of both" (Weber 1968, 54). In his analysis, people ascribe to charisma different forms, which are performed and situational. As charisma becomes routinized, qualities of exceptionalism become increasingly important, invoked through everyday practice. Drawing upon Weber's ideas about charisma, I argue in this chapter that how scientists present some information to the public in effect creates charismatic data. The way scientists treat some data creates an aura of exceptionalism and suggests that the knowledge they are presenting has a particular power. The "surprise" of charismatic data is a regular aspect of scientific presentations.

The relationship between scientific data and the person who presents them is an admixture of this power. The spokesperson's ability to present the information as an expert also gives him or her access to alliances with other experts. This phenomenon is easily observable in the circulation of images of global climate change and the ways Antarctic data and scientists have involved themselves in public discussions about climate change.

Climate change is always addressed in Antarctic public presentations, whether or not the presenter is a climate change expert. In the dozens of presentations I attended and recorded while I lived in New Zealand, every presenter referred to climate change and some of the most convincing evidence for it. Discovering and reporting on charismatic data about climate change has become one of the most popular justifications for expensive and logistically complicated national Antarctic research programs.

Climate change has come to fairly saturate popular culture and international politics, in no small part due to former vice-president Al Gore's work to publicize climate change. His documentary film *An Inconvenient Truth* (2006) won an Academy Award, and because of his work on climate change, he was awarded the Nobel Peace Prize in 2007 jointly with the IPCC. Journalists, scientists, and officials have recrafted Antarctica to be an emblem of climate change. This is not the first time the Antarctic has been the symbol of global environmental change: the continent sits underneath the hole in the ozone, an indicator of human-caused atmospheric degradation that scientists discovered in the 1980s. Recently, images of massive icebergs "the size of Manhattan" have been filmed and photographed as they floated away from the continent. Without context, spectators could interpret these images as evidence that the continent's ice is melting and breaking up, thus contributing to sea level rise. However, ice floating over water does not make the sea level rise: only ice flowing off from the continent proper can add water to the oceans. In public presentations, Antarctic scientists may address these popular images as they relate to their research and may present themselves as experts on the Antarctic in debates about climate change.

Antarctic scientists, particularly those who give presentations to a wider, more general audience or to experts outside their particular disciplines, use evidence to prove that the climate is changing and sometimes insert policy (and political) suggestions into their presentations. The work of scientists is full of opportunities to become enmeshed in, critical of, and powerful to broader cultures. Those who present information about the Antarctic—academics of diverse scientific backgrounds, bureaucrats from various national Antarctic programs, and NGO representatives—are able to and are expected to talk about climate change science and its relationship to the Antarctic, regardless of their specialties and training. Although scientists are expected to distinguish science from politics, this division sometimes crumbles (Gieryn 1995). Public presentations about the Antarctic illuminate how scientific observation becomes transformed into convincing data and how data are shaped into scientific fact and made politically meaningful in relation to the expectations of audiences. In this chapter, I focus on a suite of climate change data and the discursive and material contexts of lectures given by Antarctic experts about this data to non-expert audiences. I will analyze the process presenters create of moving from observation to data to scientific fact to political meaning. First, I will describe the concepts and uses of "beautiful" and charismatic data. Then, I will interrogate the epistemic community that groupings of data and people create in the context of Knorr Cetina's idea of "epistemic agency." Finally, I will discuss translations of data into facts and meanings to show how particular engagements with compelling, charismatic data situate Antarctic experts as spokespeople for climate change. People use charismatic data to work together, particularly in identifying each other as experts, spokespeople, and allies (Star and Griesemer 1989). Some scientists who present climate change data as charismatic do so as a way of creating epistemic relationships with the public, to make their audiences see that data as politically salient and as the basis for political action. This chapter examines the relationships among scientists, policy makers, and data that affect climate change. Expert spokespeople forge political networks with their audiences, each other, and their data in order to convince people to take political and technical action to mitigate future climate change.

Histories of Climate Change

Global prognostications and projections depict a future of an anthropogenically warmed planet. Information about climate change is presented as a consequence of modern industrial life in press releases from the Intergovernmental Panel on Climate Change or in sleek PowerPoint presentations. The contributions of a few earlier scientists, however, suggest a deeper time scale for this current moment of climate research. Svante Arrhenius, a Swedish chemist, published an article titled "On the Influence of Carbonic Acid in the Air upon the Temperature of the Ground" in the *Philosophical Magazine and Journal of Science* in 1896. He asked "Is the mean temperature on the ground in any way influenced by the presence of heat-absorbing gases in the atmosphere?" (237). He postulated that "carbonic acid"—what we now call carbon dioxide (CO_2)—in the atmosphere contributes to warmer temperatures on the earth's surface. Arrhenius is the first scientist who is known to have linked the presence of carbon dioxide in the atmosphere with global warming.

Antarctica plays a complex role in the history of climate change. It is a laboratory where climate change is depicted in real time, as seen with the B15 iceberg, and over millennia, as ice cores are extracted and analyzed to read the history of the earth's atmosphere. The continent is also depicted as vulnerable: as mentioned above, the Antarctic Peninsula is among the most rapidly warming locations on the planet. The Arctic, the Antarctic, high elevations (such as the shrinking glaciers of Mount Kilimanjaro), and low-lying islands are considered bellwether geographies that will experience climate change before areas in more temperate climate zones and on larger continents. The data extracted from these areas will help explain how the climate may change on the rest of the planet and how the political and symbolic meanings of these places may serve as a call to action. The Antarctic continent and the surrounding ocean are also often described as a driver of world climate systems, helping propel and regulate normal climate patterns for the entire planet. The continent influences the climate models the IPCC presents in its reports, which are the most influential documents for public international climate change policy.

John Mercer, an Ohio State University glaciologist, was the first scientist to discuss anthropogenic climate change in the Antarctic. His 1978 article in *Nature*, "West Antarctic Ice Sheet and CO_2 Greenhouse Effect: A Threat of Disaster," now appears prescient. Mercer used climate modeling to predict that the West Antarctic ice sheet could melt rapidly, causing sea level rise, as a result of humans burning fossil fuels. What Mercer called the "threat of disaster" continues to loom in scientific inquiries and public presentations about the Antarctic. Overtones of impending disaster help make climate change research compelling by intertwining the fate of humanity with data gathered in Antarctica.

The histories of climate change also rely on multiple temporal shifts that can obfuscate the stories and cautionary tales being told: experts who give public presentations often spend time spelling out the various geologic time scales they are working with, some of which are on the order of hundreds of thousands of years. This deep sense of earthly time frames humanity in a way that displaces us while making our current atmospheric contributions even more significant. In his discussion of the history of early geology books, Robert Macfarlane (2003) eloquently writes that "geology makes explicit challenges to our understanding of time. It giddies our sense of the here-and-now" (43). The history of the scientific study of climate change spans only a little over a century, and the hole in the ozone layer was not discovered until the 1980s. Policy makers, though, use a different sort of time scale that projects scientific models and the effects of policy actions into the future. Expert presenters juggle these time scales in their talks and look for hooks that help make sense of these complicated stories to their audiences. Using charismatic and beautiful data helps smooth out the incommensurate time scales between science, policy, and history and creates a sense that action is urgently needed.

Beautiful Data: The Keeling Curve

The problem of identifying how humans should respond to climate change can be seen as a problem of strict divisions between the sciences and politics: working on this problem relies on translations that are recognizable to both scientists and policy makers. The Climate Change and Governance Conference held in Wellington, New Zealand, in March 2006 both reified this division and strove to adequately frame the problem and think up possible climate change solutions. The conference was structured to attend to both science and policy perspectives on climate change. It took place over two days, with one day given to scientists and one day given to policy makers. In the conference program, the first day, which was labeled "Evidence from Recent Science," was devoted to scientific findings as presented by scientists.

The second day, which was labeled "Accelerating the Policy Response," was devoted to policy makers who discussed policy scenarios and a debate among participants about what can or should be done in response to the problem of climate change that the scientists had mapped out on the first day.[2] The structure of the conference, then, implied that scientists are the spokespeople for facts and policy makers are the spokespeople for action.

The conference organizers had invited experts whom they hoped would lay out the most recent information that convincingly showed that the climate is changing. On the "Evidence from Recent Science" day, a climatologist presented a graph of the Keeling Curve, referring to it as "probably one of the most famous geophysical measurements." Among climate scientists, the Keeling Curve—also referred to in its ungraphed form as "the Mauna Loa data"—is considered to be beautiful because the measurements, taken in Hawaii by the late Charles David Keeling of Scripps Institute, elegantly show an increase in carbon dioxide levels in the atmosphere over the last half-century or so.[3] The squiggles that neatly jolt up and down in a line that is clearly rising represent the seasonality of atmospheric carbon dioxide levels: it rises in the summer and drops off a bit in the winter. This graphed curve is the kind of data many scientists hope to collect: it strongly and cleanly depicts a phenomenon that has been measured carefully and with exacting standards over time. Its predictability and clarity make it the kind of beautiful data that scientists want to observe.[4] In general, scientists define beauty as that which is obvious, elegant, simple, and clear.

Beautiful data can compel and convince scientists and policy makers to act. For example, in the long negotiations over whether to build the hydrogen bomb, J. Robert Oppenheimer initially supported the idea but then changed his mind because of the bomb's destructive potential. According to his later retelling, he came to support its development again with the proposal of the Teller-Ulam design, which solved an ignition-fuel problem by creating a staged implosion. Galison and Bernstein (1989) write of his conversion: "Oppenheimer explained during his loyalty security hearing of 1954 that 'when you see something' like the Teller-Ulam configuration 'that is technically sweet, you go ahead and do it and you argue about what to do about it only after you have had your technical success'" (324). The data about the Teller-Ulam design was so compelling that Oppenheimer suspended his ethical opposition to the H-bomb. It is worth noting that, in this example, Oppenheimer presented himself as a scientist—and emphatically not a politician—in the face of hot political criticism.

Elegant data portrayed visually is also a metaphoric trope. For example, McCloskey (1985), wrote of the ideological underpinnings of a 1957 Robert Solow essay that suggested rising American incomes during that decade. She wrote of the presumed upward trajectory that "the jumble of responsibility, habit, conflict, ambition, intrigue, and ceremony that is our working life is supposed to be similar to a chalked curve on a blackboard" (1985, 84). Graphed data, in this case, can give shape to the stories we tell about ourselves and our visions of the future. Data visualization can disrupt or reinforce broader political narratives.

In addition to looking like good data should, the Keeling Curve serves as condensed visual proof that atmospheric carbon dioxide levels are rising. The elegant upward curve indicates a steep rise in the presence of CO_2 in the atmosphere. As the most numerically significant greenhouse gas in the atmosphere, the CO_2 compound often serves as the representative of all increasing greenhouse gases. When compared with other graphs of longer time scales, the Keeling Curve provides evidence of an increase in CO_2 levels that is disproportionate to any rises that have occurred in the last hundreds of thousands of years. This is caused by anthropogenic contributions of carbon dioxide in recent centuries (IPCC 2014).

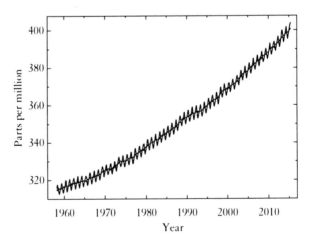

Figure 8. The Keeling Curve shows atmospheric carbon dioxide levels at Mauna Loa Observatory rising over time. Pieter Tans, NOAA/ESRL, and Ralph Keeling, Scripps Institution of Oceanography.

While advocates of climate change say that the link between anthropogenic CO_2 emissions and large CO_2 measurements in the atmosphere is obvious, a small number of politically powerful people have disagreed. Because of this contention, turning human-induced climate change into a well-defined global problem to be solved was quite difficult until recently. In addition to obtaining accurate data through meticulous experiments and producing an elegant graph, scientists had to translate this knowledge to make it significant and credible enough to become meaningful in a transnational political arena.[5]

The Wellington conference presenters tried to translate facts into policy ideas. The scientist speakers saw themselves as the stage setters for policy actions that would be determined at a later date by another group of people. However, much of the policy and political work was done within these scientific presentations as presenters repeatedly supported the strong consensus on the climate data as they discussed it. For example, after the climatologist had discussed the Keeling Curve, other panelists referenced and supported it several times. The Mauna Loa data, which has been replicated not only in Hawaii and New Zealand but also in several other research centers throughout the world, were powerful in their simplicity and their ability to be reproduced. To those trying to frame the problem, these data served as compelling evidence of global, human-induced climate change. But there is nothing intrinsically simple about this graph at all. That this curve is taken to be self-explanatory is a sign of the production of an epistemic community, in which experts share general sets of knowledge.

The policy makers who spoke on the second day outlined plans for solutions to rising CO_2 levels at local, national, and international levels. At this conference, the Keeling Curve was treated as data that bolstered arguments for the relatively accepted strategy of limiting carbon dioxide output to 550 parts per million.[6] This number works as an anchoring device for policy—the technical number conceals the political work involved in choosing that number and achieving general acceptance of it (Van der Sluijs, van Eijndhoven, Shackley, and Wynne 1998). The focus on the second day of the conference was how to build legislation and encourage action to attain this goal. The topics that were discussed included a New Zealand–wide carbon tax, reducing methane emissions among the nation's numerous sheep herds, insulating houses, the newest IPCC report, and public health issues in relation to extreme heat or other weather events.

The policy people avoided using science as the basis for their suggestions, often saying things such as "that isn't my expertise" or "I am not an expert." Science presenters would also say "this is not my area," for example when a glaciologist was describing atmospheric science, but they were much more likely to try to explain the science anyway, even if it was not in their field. The scientists seemed more vocal and competent. The policy makers were often passionate or angry about specific policy events, such as the recent failure to pass a carbon tax in New Zealand. Scientists had opinions about a broader range of issues that included their own science, other people's science, and policy ideas. The policy people took certain scientific data as fact and did not question it. Using the compelling Keeling Curve as shorthand for obvious and rapidly increasing anthropogenic climate change helped ally the science and policy experts at the conference. At this conference, the charismatic data worked as a symbol for shared and agreed-upon knowledge.

Climate change is framed as a problem that can be solved by different groups of experts and different sets of credible knowledge. Thus, the global environmental problem contains the problem of different kinds of expertise. These differences are often borne out in ethnographic depictions. At the Wellington conference, scientists and policy makers grappled with these differences in an attempt to understand and mitigate an impending global environmental problem. This example highlights the importance of careful and repeatable scientific data that can lead to clear definition of a problem, strong activism among domestic and international policy makers, and the public and political will to solve this problem. The motivation to do so comes in part from scientists' successful performance of expert knowledge before a public who have criteria for what constitutes good knowledge making and expertise. Although the Wellington conference was planned to separate science and policy—and it did so with some success—having convincing data that was agreed upon helped lead to some consensus among the presenters and attendees. The translation work done at the intersections of science and policy is particularly salient in discussions of climate change problems.

The provided and (for the most part) accepted strategy for curbing the effects of anthropogenic climate change is not a list of actions and reductions. Instead, it is found in another graph that represents another set of data. This graph extends the Keeling Curve into the future and is also often used in presentations about climate change modeling and future scenarios. Each line in the graph in Figure 9 represents a proposed policy and suggests the levels of carbon dioxide concentrations that will result if that policy is followed.

Some scientists, policy makers, and environmentalists base their arguments on the idea that CO_2 measurements in the future atmosphere should

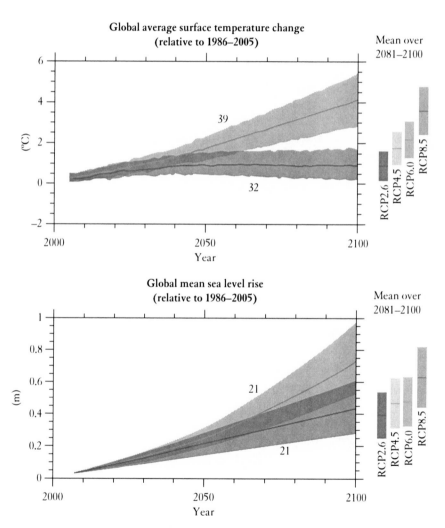

Figure 9. Representative concentration pathways. IPCC 2014.

not exceed 550 parts per million by volume. Some modelers show that this will result in some global warming and some changes in global climate, but that it will not, in their view, be catastrophic. This number, and the accompanying measurements that link the success or failure of a global human population to attempts to achieve it, becomes both the scientific data and policy options of climate change, graphed in a single image. The number also makes climate change policy technocratic, as policy makers are presented with an account of possible environmental futures based on their decisions.

Although the conference attendees and organizers maintained episte-mological divisions between policy and science, they at least managed to agree on the terms of framing the problem and a workable solution. This solution requires policy makers to legislate technoscientific practices to curb CO_2 emissions. Policy makers need to be experts at translating scientific data and advice into workable legislation and policy. The strong scientific consensus on anthropogenic climate change and rising global carbon dioxide levels forged the necessary links between political conviction and precision, beauty, and inevitability. The convincing work that scientists have done with the data has been carried forward into public presentations.

Larsen B in Public Presentations about the Antarctic

In many of the national Antarctic programs, scientists and program employees are required to present their work to the public, since the programs are sponsored by taxpayers and are presented as things that are done for the public good. "The public" can vary widely in Christchurch, where my fieldwork was based. Audiences for these talks usually consisted of several faculty members from the University of Canterbury and nearby Lincoln University, graduate students conducting Antarctic research, Antarctica New Zealand employees, former members of the Graduate Certificate in Antarctic Studies program who live locally, members of the Antarctic Society, local people who have visited the Antarctic as tourists, and other people with an interest in the Antarctic. These talks were given in lecture halls on university campuses, at city halls, and at museums. The Antarctic experts always used PowerPoint presentations: I never observed a climate change presentation that did not include this artifact of expertise.

Employees of government agencies often had PowerPoint templates that advertise their agency throughout the presentation, with insignias that appeared in the corners of their slides and were prominently displayed on the introductory and concluding slides. Academics usually did not have PowerPoint templates but would present insignias that acknowledged their institutions, funding agencies, and projects in their introductory slides or at appropriate moments in their presentations. PowerPoint slides usually outlined the main points of the speaker's presentation and served as a showcase for displaying scientific charts, maps, graphs, animations, and photographs. The scientific data provided in these slides are examples of how public presentations rely on charismatic data

PowerPoint presentations have been heavily criticized as modes of conveying information. Edward Tufte (2006), who in his book *Beautiful Evidence* writes that "the point of evidence displays is to assist the thinking of producer and consumer alike," argues that PowerPoint presentations produce what he calls "pitch culture," in which massive amounts of data are streamlined onto a single slide and written claims are shortened to soundbites (9, 166). Each slide, which has space for only a few words, Tufte argues, excludes more nuanced information. PowerPoint presentations on climate change in the Antarctic are good examples of the things Tufte criticizes. However, such presentations also include a verbal narrative. My analysis draws on the presentations I recorded. What the presenters said about the short lists of facts and the charts, graphs, and photographs they organized on their slides revealed their political and emotional relationships to the data they spoke of.

The most compelling piece of Antarctic climate change data is the charismatic example of the collapse of the Larsen B Ice Shelf. While graphs and figures depict information about extremely long-term climate change or an extremely huge piece of land, the Larsen B collapse provides information about a local, recent event. Scientists had monitored Larsen B on the ground and by satellite imagery for decades. Over the course of six weeks in 2002, it broke up and drifted in little bits out to sea. This event triggered even more collapse, because the ice shelf had been acting as a "plug" for the glacier behind it.

The satellite images of this collapse are dramatic. Presenters show photographs taken within weeks of each other, showing the complete disintegration of the ice shelf. Antarctic presenters pointed to this event as an example

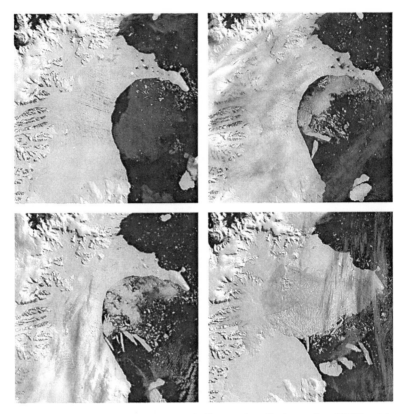

Figure 10. The breakup of the Larsen B Ice Shelf. MODIS satellite images from NASA and the National Snow and Ice Data Center, University of Colorado Boulder.

of how the Antarctic Peninsula is warming, stating that it is already exhibiting catastrophic, dramatic, and rapid climate change.

Images of the Larsen B Ice Shelf collapse echo on a larger scale the doomsday imagery that often appears in popular media accounts of icebergs crashing into a gray sea. However, experts who presented information about the Larsen B event did narrative work to demonstrate to their audiences why this case is scientifically relevant. One called the pictures he was showing a "very compelling set of images." He explained that

> this is a sequence of shots of the Larsen B Ice Shelf which is on the Antarctic Peninsula. You can already see that the ice shelf is in trouble. It [is] not supposed to have black spots on it. This is sort of a case of measles, as it

were. . . . These are actually meltwater ponds forming on the surface of this 1- to 200-meter-thick ice shelf. Over the course of the next two weeks the MODIS instrument recorded the collapse of about 40% of the ice shelf, an area of about 2,500 square kilometers, so a pretty large area. And over the last five years about 60% of this whole ice shelf has disintegrated in an irreversible fashion. Interestingly, the glaciers feeding this ice shelf have sped up at a factor of 3 to 8 once this stopper was removed from the bottle. . . . Incidentally, this summer of 2002 was very warm in the Antarctic Peninsula, so we have a direct link between a warm summer, the meltwater ponds forming, and something irreversible happening.

As the audience looked at the graphic disintegration of the ice shelf, this speaker couched his scientific narration in terms of illness. He also emphasized the specific measurements a specialized piece of technology had produced. While the MODIS satellite had provided an objective view of what was happening with the ice shelf, the expert presenter encouraged a particular interpretation of the neutral satellite images with statistics about the swiftness and severity of the disintegration of the ice shelf. He presented himself as an expert on these data as he recited the numbers and narrated the spectacle of the Larsen B collapse. Another lecturer explained his images as dramatic scenes of destruction:

> Look at this fantastic image of the Antarctic Peninsula. . . . These are big ice shelves that are already floating on the sea on the eastern side there. So some of these ice shelves have dramatically collapsed, as, again, as many of you will be aware, on the Antarctic Peninsula. There they are collapsing before your very eyes. . . . As I say, we know that temperatures have risen in this part of the world. Also, those collapses happened to correspond with summer temperatures that were particularly warm. So this shows 1986. You can see some of the fine detail where the ice was connected into James Ross Island here, go past one of those collapses, [referencing next image:] 1995, the ice is broken out, big bite out of the ice shelf, and again in 2002. This was headline news at the time. . . . These ice shelves, they are very sensitive indicators of climate change.

This speaker presents the dramatic spectacle of the collapse as an example of climate change. Like the Keeling Curve drawn from the Mauna Loa data, the images help make the data clear and compelling and the implications seem to be obvious. The chain of causal explanations this presenter described made the images do political work.

The public presentations of Larsen B images were co-produced between the presenter and the satellite imagery in a way that changed the interpretation of the images. Instead of presenting the visual data as an example of a spectacular collapse that took place within the confines of the Antarctic Peninsula, these presenters described them as convincing evidence of global climate change. Even though the scientists who tell this story may conduct unrelated research thousands of miles away in the Ross Sea region, the Larsen B story resonates with the fact that the public and scientists who present information about the Antarctic are expected to say something about Antarctica's relationship to climate change. The Antarctic is no longer just a place of penguins and beautiful stark landscapes in the public imagination; it is now a place that can tell stories and provide convincing scientific evidence of climate change. Antarctic scientists chose to tell a compelling story about the charismatic Larsen B data, and images of floating ice spreading out into the sea filled dozens of lecture halls and conference venues in my research locations. As scientists present to their audiences, they are translating and interpreting knowledge using charismatic examples. The drama of these data raise big, essential questions, and scientists present their interpretations in ways that make these seem obvious. These examples, in turn, help produce spokespeople for the Antarctic environment.

Spokespeople for the Antarctic

Climate change spokespeople agree on fundamental ideas. This agreement produces a type of community based on knowledge. Ludwik Fleck ([1935] 1979) calls these "thought collectives" that emerge around certain "proto-ideas," or "the developmental rudiments of modern theories" (25). Foucault ([1966] 1970) introduced the term "episteme," or scientific knowledge that is accepted as given in a specific historic timeframe. This concept has led to the concept of an epistemic community, which Peter Haas defines as

> a professional group that believes in the same cause-and-effect relationships, truth tests to assess them, and shares common values. As well as sharing an acceptance of a common body of facts, its members share a common interpretive framework, or "consensual knowledge," from which they convert

such facts, or observations, to policy-relevant conclusions. They identify problems in the same manner and process information similarly. They also share a common vocabulary, common political objectives to which such policies should be addressed, and a common network in which findings are exchanged and shared concerns are formulated. Although members of an epistemic community may be drawn from different scientific disciplines, all will share some common world view and concern about the same subject matter. (Haas 1990, 55)

Examples that point to climate change such as the Larsen B Ice Shelf collapse serve as foundational knowledge for the Antarctic climate change epistemic community.

The category of epistemic communities, however, is problematic because of its idealized and glossy sets of knowledge and principles of inclusion and exclusion. Although epistemic communities make significant contributions to scientific and political problems, this model does not work across groups of experts because of differences in types of expertise, because epistemic communities are far from egalitarian, and because these communities are embedded in political institutions. Knorr Cetina, who compares molecular biologists and high energy physicists, argues that the megaexperiments of the latter are "post-traditional communitarian structures" that most closely resemble an epistemic community as Haas defines it. This has shifted what Knorr Cetina calls epistemic agency—which she defines as "the capacity to produce knowledge"—into more collective and interdependent realms (1999, 167, 169). Conversely, Knorr Cetina argues, molecular biology is much more individualized and dependent on hierarchy. The epistemic agency in this discipline is largely attributable to the power of the laboratory leader. Knorr Cetina's definition of epistemic agency is useful because it adds nuance to the idea of who can have the capacity to produce knowledge and speak of it as a political actor.

Shifting the definition of epistemic subjects from people to include data as a subject of knowledge reveals a more complex community that includes the interpretations that necessarily occur when people—such as Antarctic climate change presenters and their audiences—interact with these pieces of data. This is not to say that such interactions are necessarily characterized by gaps in scientific method or in a scientist's objectivity, but rather that these interactions illuminate how people organize data and encourage

them to "speak" in order to encourage audience members to make political conclusions about climate change. Spokespeople interact with climate
change data as if they can do work to explain global climate change. The
interaction between the spokespeople and the data give the data agency.
In a similar way, charismatic climate change data are not charismatic in
and of themselves; they acquire charisma in relationship with the scientists
who use the data in ways that encourage that perception of them.

The ways speakers select examples that illustrate climate change in the
Antarctic, the way they arrange these pieces of data on PowerPoint slides,
and the way they describe these slides translates scientific observations conducted in Antarctica and analyzed in the field or in a laboratory into a
collection of data about climate change for the public. These presentations
are usually construed as education, not as explicit politics. As the presenters describe the examples they have chosen and project the images onto a
screen, the data and its interpretations make meaning for the audience.
Certain data—for example, the collapse of the Larsen B Ice Shelf—are
so compelling that their meanings translate as scientific proof of climate
change in Antarctica and as an example of the repercussions of climate
change on short, visible time scales. The presentation of scientifically cu-
rated measurements and satellite photos have an effect on non-experts.
The Larsen B collapse has become so much a part of a vague public imagination that it was featured in the opening sequence of the film *The Day
After Tomorrow* and is mentioned in two rock songs. Al Gore's documentary *An Inconvenient Truth*—an expert presentation on a large scale—used
Larsen B as one of its examples. The Larsen B data have epistemic agency;
they work to translate scientific facts into political meaning. In this case,
the agency conveys the need for caution and a sense of urgency, giving a
kind of concreteness to the futuristic science of climate change. When Antarctic spokespeople describe the visual evidence of Larsen B, it looks like
dramatic, world-threatening climate change should—catastrophic and
devastating. Larsen B does convincing work that more mundane climate
change data that depicts small and slow but serious shifts in atmospheric
gasses and temperatures cannot accomplish. Presenters engage with the
work of imbuing their public presentations with political meaning, usually during the conclusion of a science-based talk. These statements are
supported by the data they have chosen, their rationale for arranging their
examples in the way that they did, and public expectations that Antarctic
scientific research will produce some kind of collective outcome.

Charismatic Spokespeople

On my first day in New Zealand, I had planned to enjoy the vibrant confusion of culture shock and jet lag, one of my favorite sensations of travel. However, a Fulbright employee who was enthusiastic about my project had kindly scheduled a meeting for me with one of New Zealand's most prominent scientists, an Antarctic climate researcher.

After waiting a bit in a university lobby adorned with maps of the Antarctic and a fossilized tree trunk excavated in Antarctica, John, the scientist, warmly ushered me into his office.

It was immediately apparent that John had some things he needed to tell me, an American graduate student living in New Zealand in the last years of the George W. Bush administration, particularly about religion and management of the Antarctic environment.

John picked up a newspaper clipping resting on his windowsill. The headline claimed that George W. Bush said that he had been commanded by God to invade Afghanistan and Iraq. U.S. government decisions that are faith-based, suggested John, contrast with what he calls "the culture of science," which is evidence based, standardized, and peer reviewed. John's frustration with the U.S. official position of denying the scientific reality of anthropogenic climate change was palpable.[7]

His irritation was not limited to U.S. climate policy. Antarctica was already experiencing the effects of climate change and is a symbolic location for the future of the warming earth. Climate research in Antarctica is some of the most sophisticated and well-developed on Earth. However, according to John, none of this shows up in Antarctic environmental policy. To him, many Antarctic policies focus on small practices that minimize human impact on the Antarctica, such as waste disposal, fuel transfers, and rearranging rocks at campsites. John said that he found these policies a little pointless because they are not far-sighted enough in terms of global change. He said that he and his colleagues "don't care if the bases are tidy if the entire continent is starting to melt."

A few months later, a ship sailing from Ushuaia, Argentina, carried the Venerable Kirti Tsenshab Rinpoche, a Buddhist monk, and his attendants to the Antarctic Peninsula. He had come to bless the waters in the hope that he could stabilize the oceans to help mitigate climate change. In preparation, Tsenshab Rinpoche had created holy water that contained blessings from the Dalai Lama, water from each of the world's oceans, and water from places that Tsehshab Rinpoche had visited. His meditation center in Nelson, New Zealand, issued a statement that his voyage to Antarctica would help mediate climate change:

> It is believed that blessed water can stabilize and harmonize the weather in any place especially if the weather patterns are disturbed. Extremities of climate and terrain are found on the Antarctic continent that are found nowhere else in the world. By rectifying the weather in one place the weather all over the globe is adjusted accordingly. . . . To have all this water blessed and spread throughout the globe is indeed an incredible and practical way of helping every sentient being. (Chandrakirti Meditation Centre, 2005)

Tsenshab Rinpoche successfully conducted this blessing in Antarctica. While there, he dropped a carved statue into the ocean upon which was written names of people to be blessed, including the expedition's sponsors.

Of course, the Venerable Kirti Tsenshab Rinpoche is not the only person to use ceremonies and press releases to affect the global problem of climate change. Scientists also mobilize the corpus of climate change research they have conducted in the Antarctic to participate in debates about whether global climate change is real, how to mitigate its effects, and its other scientific and political meanings. Both John and Tsenshab Rinpoche bring charisma to their activism: John with the charisma of appropriate and balanced scientific research and Tsenshab Rinpoche with blessings and rituals. Sheila Jasanoff (2004, 5) argues that "how we understand and represent environmental problems is inescapably linked to the ways in which we choose to ameliorate or solve them." The converse is also true: how we choose to ameliorate or solve these problems (and who makes these choices) strongly influences how experts represent these problems to non-experts.

In the activism of John and Tsenshab Rinpoche, religion and climate change interact. For John, the religious overtones of the second Bush administration are anti-scientific, and he connects this with the intense attention the U.S. media gives to climate change skeptics. In his narration, climate change policy decisions cannot be made by referring to the relationships of politicians with God but must be scientifically valid and testable. In the example of Tsenshab Rinpoche's cruise to Antarctica, religion was depicted as a solution to climate change problems: spiritual belief is the focus of Rinpoche's climate change intervention. Such pragmatic relationships between religion and climate change science are not unheard of elsewhere. Susan Harding (2009) argues that Al Gore's *An Inconvenient Truth* "revoices" the narrative form of a Christian jeremiad. She writes that the film's narrative is "the first to oppose [the religious right] not merely through naysaying and critique, but through performance, performance that takes up the voice, the narrative, and the rhetorical forms currently dominated by the religious right and swerves them to other ends." While Harding notes that Gore's background is Baptist and that he spent some time attending divinity school, the leftist apocalyptic vision that Gore voices by intermingling his biography with climate change science is a recrafting of both religious and scientific charisma.

The notion that the boundaries between science and politics are not static, that at times they are strict and fierce and other times are fluid and creative, is not new in social science inquiries into science. In his 1983 essay "Give Me a Laboratory and I Will Raise the World," Bruno Latour reconfigured ideas about where science stops and politics begins. Latour claimed that Louis Pasteur broke down the walls between many elements of French society such that journalists, hygienists, and physicians were suddenly interested in what was happening to microbes in Pasteur's lab and in the bodies of the farm animals Pasteur had inoculated. Pasteur's work attracted interest because his scientific work traveled outside the laboratory and was translated into practical solutions to a social problem (146). No longer an inner sanctum, Pasteur's laboratory became a "lever" that affected the practice of agriculture worldwide (150). Similarly, in the presentations described above, engagements with data situate Antarctic experts as spokespeople for news bulletins about the impact of climate change. The lever in this case is the public presentations Antarctic scientists give, which operate between the inner sanctum of expensive, technical, and expertise-laden Antarctic research and the platforms where the public consumes and uses this data. The political sensibilities of Antarctic climate change data sweep throughout these places. In Latour's *Politics of Nature* (2004), he draws from and builds upon his legacy of interrogating the creation and hegemony of specific objects and subjects in scientific practice and discourse. He also tries to resituate these objects and subjects within different paradigms. He argues that scientists and policy makers must not only acknowledge their participation with nonhuman objects but must also enter the realm of participatory politics. This is what is happening in the Antarctic epistemic community. While the climate change data that scientists who work in the Antarctic present in talks is expected to be correct, produced with rigor and objectivity, it also sometimes takes on charismatic qualities from the framework scientists use when they offer it to the public.

How Scientists Become Policy Makers

Some Antarctic scientists who make presentations about climate change shift their attention to another audience: those who write policies in response to their research. Policy makers ask this of scientists, particularly the scientists their agencies and institutions fund. As I interviewed dozens of scientists in New Zealand and elsewhere, I found a range of people grappling with the issues related to inserting their own hopes about what the impact of their research projects might be or their desire to contribute to shaping the science strategies of their nation by identifying key areas for further research. Each scientist could talk about how their research helped contribute to understandings about climate change. When I asked them to describe these contributions, some felt that the ways their research fit into understandings of climate change was so obvious that it was barely worth mentioning. Others would answer this question with the disclaimer that although their research was not primarily about climate change, since Antarctica New Zealand had made climate change a priority, they could of course draw some connections between their work and climate change.[8] A few scientists created active and well-regarded research programs and provided "policy relevant science advice" to policy makers in ways that are useful to both scientists and policy makers.

On a rare free day during my short stay in Wellington, I scheduled interviews with policy makers who had to engage with scientists in order to make decisions. I hoped to learn about how individual scientists and specific scientific examples helped policy makers in their work. I first met with Michael, who is involved in making New Zealand's national science policy. The agency he works for determines which directions national research will take, who gets research funding and which geographic areas are funding targets, and what scientific problems are deemed to be in the national interest.

Michael's interest in Antarctic science centers on how much scientific research takes place in Antarctica and the quality of this research. To him, Antarctic science is different from other New Zealand science because those who conduct it claim that they maintain a national presence in Antarctica and are international leaders in Antarctic science. In addition to the knowledge their work produces, it meets diplomatic and political objectives for their nation. Also, he said, the New Zealand Antarctic program is one of the strongest ways New Zealand cultivates good relationships with the United States.

Michael's job involves advocating for science and for funding for it, making sure that ministers get good scientific advice, and contributing to efforts to provide structure and direction to science policy in New Zealand. He is concerned with improving what he calls the "science/policy nexus." As a government employee, he attends the annual Antarctica New Zealand conference—an event where Antarctic researchers showcase their work to Antarctica New Zealand and each other—because it is a forum for scientists and policy makers. There, he tries to develop relationships and form networks. Through these networks, he says, he can find people who can give him advice that is relevant to policy. He emphasizes that what he is looking for is *expertise*, not *perspectives*. He thus distinguishes between scientific objectivity and political subjectivity. He wants sound advice from people whose institutions and peers will vouch for their credibility, whose practices of expertise are such that colleagues judge their advice to be of high quality and apparently neutral. However, the difference between expertise and perspectives is more nuanced; each engenders the other. Similar to the ways scientists present data in specific ways that help audiences understand its implications clearly, these spokespeople can pragmatically present the perspectives that they see as inherent in the data. This cannot be successfully achieved unless the spokesperson initially qualifies him or herself as an expert. Michael's separation of expertise and perspectives creates a false dichotomy.

Michael told me that the scientists who are willing and able to provide this kind of expertise in a way that is meaningful to policy makers are few and far between. It is very difficult, he noted, to find an individual who is well respected as both a scientist and a policy advisor. Scientists who contribute too much to the policy side and focus their research efforts in ways that can to contribute to policy are sometimes considered "soft" scientists.

This is why many scientists decline to provide information to policy makers. Another reason is that providing such expertise often implies editing, condensing, or abridging their complex research, which, to some, is an undesirable, potentially inaccurate oversimplification of the science. The best scientists and policy advisors are those who can present their findings in a way that is easily digestible, who can perform the simplifications policy makers require. When I asked him who some of the best scientists/policy advisors were in New Zealand, the first person he mentioned was John, thus situating himself and John in the same relational web of Antarctic experts.

Later that day, I walked over to the Ministry of Foreign Affairs and Trade (MFAT) building to talk to some of the most senior people who represent New Zealand's interests in international Antarctic policy. This ministry is roughly equivalent to the U.S. Department of State and employs New Zealand's foreign service representatives. I had met several MFAT employees at Antarctic Treaty consultative meetings and through my research participants in New Zealand. I had contacted one of these people, Sandra Pope, to try to set up an interview. She arranged a brief meeting for me with Trevor Hughes, the head of New Zealand's delegation to the Antarctic Treaty consultative meetings, and herself.

Sandra and her colleague Jannine McCabe, who also works on Antarctic policy, sat with me in a conference room while we waited for Trevor to arrive. The three of them form the ministry's Antarctic Policy Unit. All three seemed to be incredibly busy and spoke quickly and confidently about their roles and about international and national Antarctic issues, performing another mode of expertise that felt negotiated and official and contrasted with the scientific, usually academic, performative expertise of climate scientists I knew. I had the impression that the fact that all three people from the Antarctic Policy Unit were in the room at once meant that I was being delivered a competent, detailed overview of the standard government story about Antarctic policy in New Zealand. During this three-person interview with the Antarctic Policy Unit at the Ministry of Foreign Affairs and Trade, the members exemplified the professional and bureaucratic characteristics of international policy making. Since it was two employees being interviewed along with their powerful supervisor, I did not expect to hear personal opinions or examples of times when New Zealand policy making did not succeed in Antarctica. Instead, for a few minutes, I was able to watch three people collaborate at describing how Antarctic policy in New Zealand is made.

The Antarctic Policy Unit is an example of the Ministry of Foreign Affair and Trade's realm of "multilateral environmental issues." As the head of delegation at the Antarctic Treaty consultative meetings, Trevor is the primary policy spokesperson for New Zealand on Antarctic issues. Sandra primarily works on the Antarctic Treaty meetings, or "land issues" while Jannine focuses on "sea issues," doing much of the policy work for the Commission for the Conservation of Antarctic Marine Living Resources.[9] These three policy makers told me that their explicit goals center on upholding international Antarctic law and maintaining credible institutions, both in New Zealand and in other Antarctic states. Trevor sees his key role as keeping the Antarctic Treaty System credible and relevant. Right now, he says, only a handful of members are moving the treaty forward; to him, this positions these nations as the credible Antarctic policy experts of the world. At this level of policy making, nations become the experts instead of individual people. However, the members of the MFAT's Antarctic Policy Unit tap the expertise of local scientists to bolster New Zealand's role as an expert Antarctic policy nation.

To this end, the Antarctic Policy Unit also convenes an advisory group called the Officials Antarctic Committee, which includes representatives from all government ministries that have some concern with Antarctica, broadly defined. For example, the Department of Conservation, roughly the equivalent of the U.S. National Park Service, has a member on the Officials Antarctic Committee because the Antarctic Policy Unit wants an environmental management expert on the committee, even though the Department of Conservation is not actively involved in Antarctic issues.

To prepare for the Antarctic Treaty consultative meetings, Sandra develops policy through consultations with the Officials Antarctic Committee, NGOs, representatives of relevant industries, and other stakeholders. From these conversations, the Antarctic Policy Unit forms policy views and presents them to the Members of Parliament who will decide on the positions. The members of the unit write working papers and information papers. After that, they present the papers to people in other national agencies who can provide additional expertise. Also, Sandra gathers advice relating to these papers from people who are informed about New Zealand's relationship with states that may be affected by their content.

The Officials Antarctic Committee is the primary sounding board for the Antarctic Policy Unit's policy development. Trevor is the chair of this group. The committee meets quarterly, allowing for the consistent development of policy advice. At these meetings, subgroups work on specific issues. Trevor, Sandra, and Jannine also have joint meetings with the delegation of the Committee for Environmental Protection, among other people who attend the Antarctic Treaty consultative meetings. Among other kinds of expertise (about political matters, about policy strategies, about logistics, about environmental issues), members of the Antarctic Policy Unit consider advice from scientists while they are writing policy documents. These documents and the collective and multiple forms of expertise condensed within them are the primary, formal mode by which the members of this New Zealand ministry convey their status as international experts.

At this interview, the members of the Antarctic Policy Unit referred to scientists as either "stakeholders" (as they are in consultations with the Officials Antarctic Committee or during less formal opinion gathering) or as science advisors. John was mentioned at the top of the list of trusted individuals who can supply policy makers with credible scientific advice. Implied in this statement is that his advice can translate into policy in a rather straightforward manner. In New Zealand Antarctic policy networks, John has situated himself (or has been situated) as one of the individuals who can describe scientific findings in ways that can become translated into policy outcomes.

At this high level of policy making, gathering scientific advice is seen as part of a necessary and routine process. Consulting representatives of Antarctic scientists—along with Antarctica New Zealand workers, the largest segment of the New Zealand population to spend time in the Antarctic—demonstrates responsible, democratic policy making. Much of the rich scientific detail that science advisors offer policy makers takes place in settings somewhere between informal meetings and workshops and the international policy forum. Perhaps because of the varieties of expertise, scientific information becomes translated into shorthand and code words when it enters policy texts: scientific data is transformed into terms such as "climate change" and "biosecurity." However, implicit and explicit references to scientific research and to consultations with active Antarctic researchers are expected and are part of what policy makers do to demonstrate their competence. Policy proposals that do not give attention to recent scientific research or that seem to be based on convincing evidence that is not scientific (emotive or nationalistic arguments, for example) are seen as less credible and are less likely to pass at the Antarctic Treaty consultative meetings.

An exception to the requirement that policy makers should wait to make decisions until all of the scientific data has been gathered is the precautionary principle. The precautionary principle is the scientification of doubt and risk. When early scientific evidence points to a problem such as a declining penguin population or an acceleration of the melting of glaciers, calls for policy strategies to address the problem or to slow its progress may be made. At the Antarctic Treaty consultative meetings, national delegations have different tolerances for employing the precautionary principle. For more environmentally focused states such as New Zealand, the precautionary principle is often called upon as justification for imposing stricter environmental regulations on the continent. Other delegations may respond to such calls by requesting that a policy decision be delayed by a year so more scientific data can be gathered. In Antarctic policy making, the precautionary principle is usually deployed in more local decisions such as whether to list or delist threatened and endangered Antarctic species. However, in the international debate about climate change, the debate between the camps who advocate the precautionary principle and those who advocate waiting for more information is more intense.

The precautionary principle is not only the technocratic device for dealing with doubt and risk, but it is a scientifically legitimate policy tool for reacting to charismatic data such as the fascinating pieces of ice named B15 and Larsen B. These are compelling examples that suggest future climate change scenarios on a larger scale. However, because their imagery is so compelling, charismatic events can serve as catalysts for invoking the precautionary principle on a broader scale than the data might suggest. The precautionary principle—a completely uncharismatic policy device—works in the space between data and fear. Conservativism, deferral of decisions to a future time, and the refusal of spokespeople to make grand charismatic gestures are other ways to respond to extraordinary events. Charismatic data speaks to broad audiences because they link to broad concerns, interests, and dreams. Thus, while "everyday" science advice passes though committee and meetings, subtly shaping policy, charismatic data can quickly effect policy changes because of its political messages and its ability to generate political will in public and policy settings.

Charismatic Data in Motion and Practice

In November 2006, several of the Antarctic scientists I worked with were asked by a helicopter tour company to narrate tours off the coast of Timaru, south of Christchurch. About 100 Antarctic icebergs that were rumored to be broken up bits of the famous B15 iceberg had traveled up the currents from the Antarctic to offshore New Zealand. The tour company promoted these helicopter trips as an once-in-a-lifetime chance to see a bit of Antarctica in New Zealand. The scientists–tour leaders—spokespeople for the floating remnants of B15—retold the stories about climate change, logistics, penguin adaptations, and Antarctic glaciology, weaving together scientific data, popular anecdotes, and the powerful images of drifting, out-of-place Antarctic icebergs. When the helicopter tour companies enlisted distinguished, senior Antarctic scientists to work as guides, the scientists became spokespeople for the information these pieces of ice contained, interpreted as climate change data that had migrated to New Zealand.

Charismatic data are made through the recording of facts, scientists' subjective observations, and the ways this information is presented to non-expert audiences. Data becomes charismatic when experts can produce a moral story from it that relates to human hopes and fears. It is not just the data that are charismatic, though, it is the spokespeople, too. Data and scientists form a cascading relationship where the data are charismatic and the expert presenters are heroic and self-effacing. Expert spokespeople perform these charismatic scientific stories and are responsible for educating non-experts about the data's existence. However, the data have epistemic agency—the ability to generate scientific interest and political urgency. When scientists sort out their PowerPoint presentations into discrete sets of examples of climate change in the Antarctic, they reinforce the political power of these specific bits of data using rhetorical expertise and scientific authority to produce charismatic data. The most skilled can pass this charismatic expertise on to policy makers without compromising their objective, scientific status. This is the move that makes a few people both successful scientists and successful policy makers.

Conclusion

The Technocratic Governance of Nature

The Antarctic, as a place, is an exception and an ideal. The vast wildness of Antarctica inspires adventure, exploration, scientific innovation, and international cooperation. In this arrangement, people hope that great science will turn into smart, effective environmental regulation. In short, while technocratic management elsewhere might feel like a nerdy burden, a leftover of 1950s efficiency politics, in Antarctica, technocracy is part of a utopian environmental future. Among Antarctic people, the broad consensus is that governance informed by scientific knowledge has the potential to improve human and nonhuman lives and homes.

This ethnography has tracked this idealistic impulse in the context of the management of the Antarctic environment, as shaped by a community of expert scientists. There are hiccups along the way—consensus-based decision making that is informed by scientific expertise and executed by state governmental delegates is bound to be uneven and saturated with concerns about sovereignty. As an organization whose members must expend millions of dollars for scientific research and logistical support, the

Antarctic Treaty System is shaped by global inequity. Nonetheless, the apparent blank slate of Antarctica has served as a place to practice a model of epistemic technocracy for nearly sixty years.

Antarctic scientists and policy makers, as well as their relationships with nonhuman objects, life forms, equipment, and paperwork, form models of environmental management that usually manage to be effective. The case studies I analyzed in this book suggest that Antarctic governance is an example of how scientific knowledge and expertise takes form as policy. When we consider the ways that Antarctic scientists and managers use their expert knowledge to make governance decisions and the moments of slippage and controversy that unfold in these attempts, we are able to gain a better understanding of how knowledge can form governance (and how it cannot or can only partially do so) at the points where both science and policy become approximate, symbolic, or chimeric.

In this ethnography, I have used the concept of epistemic technocracy to describe the expertise-driven form of governance that brings technology, bureaucracy, and science together. It is well known that contemporary technocratic forms of governance use expertise. However, in expertise-based governance systems such as the U.S. federal government, science can be dismissed in politics and policy. Observe the claims by climate-contrarian members of Congress that they are not scientists to both harness scientific evidence and to dismiss it in the political arena (Chait 2014). In Antarctic governance, dismissing science is not an option. Scientific research forms the core rationale for Antarctic governance. This is why I have proposed the notion of a technocracy that is governance forged by consensus about scientific endeavors. The model of epistemic technocracy, as told through ethnographic accounts of definition making, policy making, and slip-ups of implementation, does the following cultural work: 1) it produces knowledge that forms the basis for an epistemic community; 2) it helps experts function as contributors to governance; 3) it imbues science, scientists, and nature with particular cultural lives and meanings; and 4) it either incorporates or ignores cultural difference for the sake of acknowledging expertise.

Producing Antarctic Knowledge

This book tracks how knowledge is made and how this knowledge travels in the nebulous Antarctic community. Scientific data pours out of

Antarctica as the result of the efforts of hundreds of researchers and the people who support their work. Scientific findings change our understandings of the universe and nature, make careers, and open up new directions of inquiry. This book investigates how scientific knowledge is interpreted within the governance structure of the continent, how people who manage Antarctica use scientific data to inform regulatory policies, management decisions, and strategic scientific directions for their national Antarctic programs. This work is technocratic, combining imaginations about the vast Antarctic continent with the bureaucratic mundane in productive and delightful ways and sometimes in ways that are conventional, even flat-footed. During my fieldwork, I came to love how Antarcticans juggled this juxtaposition of extreme nature and extreme bureaucracy, as if one needed the other to pull Antarctic work toward some kind of comprehensible center.

This juxtaposition—of vast, unrelenting, sublime Antarctic wilderness with the mountains of paperwork, rules, and committee discussions that can continue for a decade without resolution—produces knowledge of the Antarctic as much as the serious work of Antarctic researchers and managers does. Antarctic people play with this juxtaposition, try to manage it, and try to situate their findings and ideas within the almost overwhelming superlatives of an almost unimaginable place and an unwieldy national and international regime of governance. This work occurs through attention to form, both of the Antarctic environment and of the material practices of procedures I have described in this book.

The production of knowledge about Antarctica also has a long history, one that exceeds and sometimes parallels the history of science more generally. Antarctica's existence was hypothesized millennia before humans sighted the place. For thousands of years, a speculative Antarctic offered visions of the nature and substance of our planet. Antarctica's earliest human visitors carried with them Western values about exploration, knowledge, imperialism, and resource extraction—topics that continue to be explicit and implicit parts of the fabric of Antarctic human activity.

The scientific knowledge that experts read from the Antarctic is about nature, as all scientific research is. While Antarctic life forms are often extreme, hardy, endemic, and adapted to specialized conditions, much of the concern of scientists and managers about Antarctic nature has had little to do with the uniqueness of Antarctic life and more to do with nature in the rest of the world, nature at "home." Scientists relate to nature in disciplined and nationalistic ways that are informed by the environmental, militaristic, or other characteristics of their training and their national Antarctic program. Though Antarctic nature forms the material that scientists engage and policy makers manage, the people I interviewed saw Antarctic nature as symbolic and believed that meanings about that nature were made elsewhere.

I contend that in the Antarctic, the epistemic community matters more than the literal place. The epistemic technocracy of Antarctica comes to stand in for nature, powerfully characterizing and ordering it. Although human culture exists and is produced in Antarctica, much of the cultural meaning of the place is made in the temporary meetings of the Antarctic epistemic communities that have formed around the world. That the culture of the Antarctic spreads beyond the continent says something about Antarctic nature and the temporary and piecemeal, partial ways that Antarctic people belong to the continent and to other places; the culture that Antarctic people have created can be moved without falling apart.

Governance as Expertise (and the Privilege of Experts)

At a simple level, experts are people who have expertise. Developing a reputation for expertise is a process—experts cultivate their expertise, deploy it in various forms, and shape their environments with it. This process is historical, in that occurs over lifetimes and over the longer time periods of the histories of scientific disciplines and of the accumulation and building upon of knowledge in particular epistemic communities. Expertise is also participatory; it relies on people communicating forms of authority in collaboration with others who can also signify expertise.

Collaborative expertise extends beyond anthropocentric notions. In the examples of rocks, microbes, nematodes, and ice that emerged in my interviews with scientists, nature and the data that mark the first scientific distillation of nature collaborate with humans generally and with scientists specifically to form particular understandings of the world. These understandings, translated through rigorous rituals such as having one's publications peer reviewed, obtaining funding, and getting one's samples through the biosecurity checkpoint of an airport, form experts and expertise.

This form of governance—one that is run by and in the name of experts and an ideal of international cooperation—builds exceptions into policy that are both informal and formal. Most notably, epistemic technocrats make exceptions about environmental management policies for some scientists. To be sure, technocrats scrutinize these practices through their review of environmental assessment paperwork. The experts who participate in the assessment process take this work seriously—this is no rote rubber-stamping process. Exceptions will not be granted when there is another viable path forward. However, typically, policies do not stand in the way of groundbreaking research.

That exceptions are seen to be given as a matter of course speaks to the trust, authority, and sound judgments experts are assumed to have. Antarctic scientists, for example, have all been trained in scientific research methods and know how to address and perform environmental and other ethical types of scrutiny. National Antarctic programs and the peer review processes they enact ensure that scrutiny of a research project's methods and intellectual soundness takes place at least a year before a research team physically travels to the ice. Both authority and certification are evident in Antarctic research, and the discursive (and often material and practiced) allegiance to environmental assessment is a core value of Antarctic scientists and policy makers.

The Ritual of the Process

Policy and practice—and the oversimplified classifications of discourse and materiality—are created in ways that illuminate the interdependence of these concepts as they are enacted and used in the world. Policy is what is created in meetings, documents, and structured rules and regulations. Practice, particularly for scientists, is bound up in the imagery of "the field" or "the lab" and in tactile engagements with equipment and the subjects and objects of study. If we refuse the easy division of discourse and materiality, it becomes obvious that the translations of policy and practice as they are moved around in Antarctic environmental management are parts of the same system. This system of policy and practice informs and does work on and with procedures and documents.

Both policy and practice rely on procedure. Developing, carrying out, and tweaking procedures are part of the work that policy makers and scientists do. The ordering of life and work, especially the steps people should take to conduct work that is mindful of the need to protect the Antarctic environment, relies on agreement about procedure. Procedures are discussed, debated, written down, published, and distributed, but they are also enacted and adapted in the field. The unorthodox translations of procedures are not devious acts but a continued process of editing that engages with the material and social conditions of the idealized procedure.

Documents are the material products that begin to spell out procedures. The milieu of Antarctic environmental management is chock-full of electronic and paper documents—policies, agendas, environmental impact statements, articles in journals, first aid manuals, memorabilia, personal journals, press releases, data stores, permits, and on and on. These documents are more than the material remnants of negotiated discourse: the ritualized practice of producing these documents contains traces of the encounters that haphazardly emerge through the contingent moments and movements Antarctic people and things make. The documents that are produced contain traces of the practices of relationality, ethics, and other motivations.

While Antarctic material life and nature figure significantly in these documents, the process itself becomes the main signifier. Meaning is made by following the process or by working around and through the process while simultaneously upholding other processes. The success of Antarctic environmental management, it sometimes seems, is determined by the ability to keep an idea on the agenda long enough for it to become "mature," or accepted as consensus as a matter of course. The work of everyday Antarctic policymakers who move ideas through the policy system gives shape to Antarctic futures. The apparent reasonableness of this work itself forms authoritative power.

Charisma: The Cultural Lives of Science, Scientists, and Nature

Another factor in the epistemic and technocratic relationship of policy and practice is consideration of an audience, and specifically for Antarctic environmental management, a public audience. For example, Antarctic scientists who are funded by the U.S. National Science Foundation are expected to contribute to national knowledge. NSF grants require applicants to write about how their research will contribute to the public domain, broadly defined. This requirement has led to a proliferation of expedition blogs, websites, public presentations, curricula, and a variety of other attempts to connect a vague, interested public to Antarctic scientific research. Many other national Antarctic programs have similar requirements that their scientists present some of their work to "the public."

International Antarctic policy—particularly in the form of the Antarctic Treaty consultative meetings—also provides information to the public, even as it effectively limits public participation. Host cities and countries of the ATCM always have a series of public events, such as art exhibitions, documentary films, science talks, and children's activities in correlation with the meetings. Each ATCM also has a small press office at the meeting venue where participants can file press releases and journalists can find insiders to speak with for their reports. Media representatives are usually allowed into the meeting proper for only a very short time (about one hour) during the opening of the meetings. No one without a credential is admitted to the ATCM.

The opening and closing of Antarctic communities, places, and work to the public is not just about access to information, decisions, and ideas about the continents. It is also about who these imagined publics are that must be allowed and disallowed. In both the instances of providing information to and withholding information from the public, there is a heavy discourse about access to the continent, the assumption of certain imagined fascinations with Antarctica, and the idea that fragile, diplomatic international consensus cannot withstand democratic or otherwise participatory input from the public. The boundaries and the access points they provide to Antarctic science and policy are attempts to define and manage outsider public audiences. These audiences want to know a bit about taxpayer-funded research and what living conditions are like at research stations and field camps. The blogs and websites Antarctic scientists maintain suggest that the public expects reproductions of heroic exploration narratives, images of spectacular nature, and a sense of the animals, plants, and rock and ice formations that populate Antarctica. They want brief news updates on how governments and scientists are making decisions and doing new things on the continent. However, political participation by the public is tricky at best and impossible in the most exclusive spaces of the ATCM. Too many opinions and interventions could obfuscate the already difficult and fragile work of establishing complete consensus. While public audiences are expected to be informed and intrigued, they are not meant to have access to Antarctica itself or its government. Though tourism increasingly troubles the notion of a closed, protected continent, insider Antarctic scientists and policy makers reify the continent as a place of peace, science, and environmentalism. This excludes those with other interests, those who harbor a range of discordant opinions, and even an array of partial insiders whose access is mediated by politically powerful gatekeepers. The audiences for Antarctic policy and science consist of a carefully screened and managed public that does not have access to the continent or to meaningful influence on Antarctica's political futures.

The Antarctic spaces that open up, in contrast, rely on forms of charisma to do convincing work of various sorts or to speak for nature in the broadest sense. Antarctic scientists invoke and cultivate charisma on behalf of themselves as noble explorers and on behalf of their data as indicators of the future of the earth. Charisma becomes routine, expected, performative, and part of the power of epistemic technocracy. Charismatic landscapes, people, and data provide the public and transformative levers of everyday governance.

Expertise and Cultural Difference

Cultural translations of environmental assessment, sound science, and environmental management are not uniform, although the apparently neutral language of scientific practice may make it seem so. Both the translations and apparent neutrality are epistemic and technocratic characteristics: both tend to bolster majority and/or mainstream perspectives while sidelining or erasing those that are marked with difference.

Cultures of expertise everywhere are both performative and process-based. Experts are only ever partially Antarctican, and this book describes moments when performing multiple cultures in Antarctic arenas can bolster or weaken perceptions of an individual's expertise.

Not many people are skilled at negotiating cultural difference in the epistemic technocracy of Antarctica. In the examples in this ethnography, cultural differences were usually smoothed over in process or ignored as intransigence. The fact that both international policy and Western science require adherence to particular practices usually manages to silence alternative perspectives and models. Because the Antarctic Treaty System claims to manage a place of peace and science, erasures of certain knowledges and understandings of nature means that peace comes at a high cost.

In Conclusion

This book illuminates some of the hopeful interventions that people conducting science and making policy for Antarctica work to make. Much of this work hinges on the powerful concept of the Antarctic as a geographic frontier on planet Earth. In this frontier, the territorial and nationalistic work of Antarctic governance fuses with the apparently disinterested work of scientific research. The Antarctic represents a frontier that potentially can be claimed and knowledge frontiers that offer the potential of expanding scientific understandings of the world. An example of the latter is the cutting-edge work on climate change that is being done in the Antarctic as the edge of the continent melts.

Many of the policy makers and scientists I have worked with are idealistic and innovative. They see Antarctica as an environmental "last wilderness" where standards are high and where a policy-making climate supports international cooperation in a somewhat odd national milieu (with no national space, for example, and where unusual border sharers live next door to each other at field camps). The people who go to Antarctica also are usually Antarctic experts: few novices manage to wander down to the ice without training and preparation. Thus, the people who visit and work in Antarctica are experts who are knowledgeable about the policy decisions their nations agree to and enact these policy decisions. There is a willingness among most people who travel to the Antarctic to conduct themselves differently on the ice in order to protect it: they engage in practices that would seem absurd or extreme in almost any other part of the world. However, they do not implement Antarctic rules and regulations at simple face value; they interpret and shape them and therefore make them. Science, too, translates through this policy system, and scientists manage their environment through both their research work and their daily lives in Antarctica.

Notes

Introduction

1. While science studies scholars often focus on enactments of expertise, it is important to note that expertise is also performed in ways that makes it legible cross-culturally. Expertise is an activity, a mode of becoming, not a mantle that experts wear (Carr 2010). Antarctic experts must cultivate and sustain their expertise and implement and mobilize it for specific audiences. I will demonstrate the efforts Antarcticans make to make their scientific expertise legible to, even inherent within, international environmental policy.

2. That engagement with the objects of expertise helps make experts (Lambek 1993) and that expertise can be corporeal if not embodied (Boyer 2005) is a result of cultural learning, a kind of epistemological disposition that one is trained into in expert life. Goodwin (1994) grounds this approach with what he calls "professional vision," an inclusive paradigm by which experts' interactions with things, each other, and broad discursive frameworks allow them to understand their objects. This understanding involves serious perception, classification, and interpretation. How many people can hold many nuanced details in their minds at once and hold a high-level conversation about these details? Typically, the answer is not many, so governing through expertise involves shorthand, trust, communication, delegation, and an explanation of uncertainty in order to make any headway.

3. Ferguson (1990) makes a similar formulation in his analysis of development failures that produce development.

4. Christchurch is a picturesque town with deliberate echoes of an earlier England. Some visitors and residents of Christchurch call the city "more English than England." However, the devastating and deadly earthquakes of February and June of 2011 profoundly impacted the city. Some

of the structures I mention below were destroyed and have yet to be rebuilt, the University of Canterbury's enrollment has dropped, and many residents, including some of my friends and collaborators, have relocated. City planners are designing a modern, green city.

5. In my time there, New Zealanders often described their nation's policies and practices with pride, including strong bicultural practices between indigenous Maori and pakeha (of European ancestry) people, anti-nuclear laws, environmental regulations, secularism, anti-war stances, and queer rights activism.

1. The Imagined Antarctic

1. Steinberg, Tasch, and Gerhardt (2015) write about environmental imaginaries at the other extreme of the planet in *Contesting the Arctic: Politics and Imaginaries in the Circumpolar North.* While their findings and mine have significant conceptual overlap, we differ in our analysis of geography, culture, and the environment.

2. Symmes's quest is depicted on websites devoted to the history of science and alternative science, such as Lee Krystak's webpage "The Hollow Earth," accessed April 14, 2008, unmuseum. mus.pa.us/hollow.htm.

3. The PBS documentary states that this ad appeared in *The Times* of London, but a copy of the ad has not been located.

4. "Shackleton's Voyage of Endurance," an episode of *NOVA,* originally aired on March 26, 2002.

5. D. Graham Burnett's massive tome *The Sounding of the Whale* (2012) provides a comprehensive analysis of how Antarctic and sub-Antarctic whaling became scientific and influenced "rational" ecosystem-based fisheries. The Convention on the Conservation of Marine Living Resources (CCAMLR) is an international regime that is deeply entrenched in the logics of rational-use fisheries. The dramas of contemporary Antarctic whaling (i.e., those of the Sea Shepherd Conservation Society) can also be analyzed in terms of science, resource, the Cold War, and national sovereignty that both Burnett and I think about (though not always in sympathy with each other) in our respective research. Whaling is a critical part of the history of Antarctica, but since I chose to focus my ethnographic research on land-based science, policies, and management, the Southern Ocean and the animals who live there figure less prominently in my ethnographic landscape. My choice is an extension of how the Antarctic Treaty System divides its policy attention: terrestrial matters are articulated at the ATCM and matters related to marine life are discussed at the meetings of the CCAMLR.

6. Deep oceans are often considered to be the other correlate of outer space on Earth.

7. Molodeznaya, Druznaya, Leningradskaya, Russkaya, and Soyuz are inactive Russian/ Soviet bases in the Antarctic.

8. National Antarctic programs were engaged in much preparation for the International Polar Year during my fieldwork in 2005–2006. Governments earmarked extra money to support spectacular scientific projects at both poles.

9. The Chilean base Villa las Estrellas on King George Island also houses families with children.

10. See Brosius 1997 for a brilliant analysis of such discourse.

11. Some nations have more resources than others to devote to environmental management education for their scientists and base employees. While all attempt to comply in spirit with the dictates of the Madrid Protocol, there is a distinct possibility that the types of training and practices I observed and participated in during my trip south with Antarctica New Zealand would differ from other countries' approaches to implementing the Madrid Protocol.

12. The book *Big Dead Place* (named after Johnson's now-defunct website of the same name) provides satirical insight in to what living and working at McMurdo Station as an employee of Raytheon entails. (Raytheon contracted with the National Science Foundation to run the base.)

2. The Environmental History of the Antarctic

1. These recollections derive from ethnographic interviews I conducted.

2. I define "environmental impacts" very broadly, as any human influence on the environment, any human trace in Antarctica that can be noticed, visibly, chemically, microscopically, etc. The task of choosing which impacts to track and minimize is made very difficult because of the vagueness of "impacts" as a category.

3. Some people involved in management of the Antarctic environment acknowledge that some national Antarctic programs have different readings of expected environmental practices on the continent and surrounding waters. Some acknowledge that some national Antarctic programs have limited budgetary and staff resources to ensure that environmental practices are followed to the letter. Not all national bases have had the same success implementing all aspects of the Madrid Protocol, even though all parties have agreed to the document.

4. The precautionary principle suggests creating policy to mitigate potentially high-risk problems before these problems have actually been observed or documented.

5. These scientists will remain unnamed to protect the anonymity that Sam assured in the original petition he sent out.

6. Japan halted scientific whaling in the Southern Ocean in 2014, but it resumed the practice after a one-year break.

7. The other two environmental groups with expert status at the Antarctic Treaty consultative meetings are the United Nations Environment Programme and the International Union for Conservation of Nature, both international governmental organizations. The ASOC maintains a much more activist position in the treaty meetings than these two groups.

8. I attended the 2006, 2007, and 2009 Antarctic Treaty Consultative Meeting as an ASOC delegate. I had tried for several years earlier to attend the meetings with the U.S. delegation but was politely refused. I was told that the United States does not add non-essential personnel to its delegations. Finally, through some serendipitous connections I made in New Zealand, a senior ASOC delegate convinced ASOC to add me to the delegation. Being on an expert delegation instead of a consultative party delegation is an interesting experience: depending on national environmental positions, state delegates will avoid ASOC members, speak with them only in isolated places (a U.S. delegate whom I had previously met only spoke with me in the restroom, for example), or will make a point to publicly visit the ASOC delegation table, which is located behind the consultative party tables. However, this experience changed over time and with a new presidential administration in the United States.

9. I have served the U.S. Department of State and ASOC in that capacity since concluding the research for this project. No data from those meetings are included in this book.

10. The first wave of environmentalism, in this narrative, took place in the 1960s and is epitomized by Rachel Carson's *Silent Spring*.

11. See the website of the Center for Outdoor Ethics, titled Leave No Trace, at www.lnt.org.

3. Sensing the Ice

1. The Antarctic Treaty (1959) suspends the territorial claims of seven claimant states: the United Kingdom, New Zealand, Australia, Argentina, Norway, France, and Chile. The treaty also forbids any other states from making claims. This agreement was sought because of increasing interest in claims by the United States and the Soviet Union. These nine states plus Belgium, Japan, and South Africa are the original treaty signatories. More states have been added, and all consultative parties have a sustained and active scientific presence in the Antarctic. Other states without an Antarctic scientific program can sign and observe the annual Antarctic Treaty Consultative Meeting as nonconsultative parties, as can a small number of expert and observer groups.

2. While this chapter focuses on glaciologists and ice-sheet modelers, both of whom are concerned with the future of Antarctic ice, the interests of researchers in the Antarctic are diverse.

The South Pole is home to the world's largest neutrino detector, located a kilometer down in the ice. Its mission is to try to capture extraterrestrial neutrinos as they zip through our planet. Biologists and ecologists examine nonnative species, fungal communities, and the multiple species and colonies of penguins. The small amount of ice-free areas in the Antarctic are sites of intense scientific interest because of the remarkable natural processes and species found there and because of infrastructure development, as it is much easier to build on solid ground than it is on snow. The rare and unique environments in Antarctica have inspired some of the world's most creative and comprehensive environmental management strategies.

3. Ice-sheet behavior occurs independently of warming and other anthropogenic forcings. However, some parts of the Antarctic ice sheets are changing more rapidly because of warming air and oceans.

4. The names of interview participants are pseudonyms.

5. The first core into Lake Vostok was drilled in 2013. Environmental impact assessments were discussed for almost a decade preceding this activity, at various ATCM.

6. In the Antarctic Treaty System, consensus is achieved when all questions have been addressed. Sovereignty is generally upheld at all costs (O'Reilly 2011).

7. This was particularly noted several times in the United Kingdom, perhaps because of the British masculine Antarctic mythos, which was underscored by the fact that the British Antarctic Survey did not allow women scientists or workers at its research stations until 1986. No woman overwintered there until 1997. The pace of change in gender representation has been fairly rapid since the late 1990s, and a woman now directs the British Antarctic Survey.

8. The grounding line is the front wall of the glacier, usually where the foot of the glacier stops on land. The West Antarctic Ice Sheet has a marine grounding line; it is under water. This changes how the grounding line behaves, but if a grounding line retreats, it may trigger disintegration of the entire ice sheet.

4. Samples and Specimens at Antarctic Biosecurity Borders

1. This is not the first time nature has received such diligent and potentially violent security oversight: the history of the United States Army's presence in the early national parks also merged nature protection and national security (Sellars 1999).

2. This approach, of course, is in direct opposition to certain political approaches to anthropogenic climate change that insist on studying the problem completely before implementing mitigation strategies. In the example of Antarctic biosecurity, although the problem is not yet well understood, management strategies are already being devised and put into practice.

3. New Zealand's status as the location for Lord of the Rings film trilogy leads to comparisons and advertisements involving Middle Earth.

4. What was known as the Ministry of Agriculture and Fisheries is now called the Ministry for Primary Industries. In April 2012, New Zealand merged the Ministry of Agriculture and Forestry, the Ministry of Fisheries, and the New Zealand Food Safety Authority into the Ministry for Primary Industries.

5. Managing Antarctic Science in an Epistemic Technocracy

1. I first met Henry in 2003 on my exploratory field trip to New Zealand, and it was he who introduced me to the emergent Antarctic Specially Managed Areas. This early conversation sparked the initial direction of my dissertation research: I owe him a great debt for this inspiration.

2. Antarctic Treaty language often protects specific values—wilderness, aesthetic, scientific, and so on—instead of protecting "the wilderness" or "the environment."

3. Meetings are now ten days in duration.

4. Malaysia signed in 2011.

5. Information papers simply provide information. Secretariat papers are logistical papers issued by the new Antarctic Treaty Secretariat; for example, one of the secretariat papers is the meeting agenda. Working papers are more powerful: these can be turned into measures (policy). Only member states can offer working papers at the meetings. The organization I attended the ATCM with, the Antarctic and Southern Ocean Coalition, is only able to submit information papers.

6. Tectonic Time and Sacred Geographies in the Larsemann Hills

1. Since 2007, India has operated only one base, named Maitri, established in 1988. Its first base, Dakshin Gangotri, was established in 1983 but was abandoned after being buried under drifting snow.

2. This group of delegations, knowing India's interest in the area, had invited Indian representatives to join the ASMA management planning group, although India had not joined at this point.

3. Recently, claims in the Arctic based on geologic arguments have been deployed. The claim getting the most attention is Russia's claim that sea floor under the pole is an extension of its borders (Blomfield 2007).

4. There is another arc of criticism of this story that this book will not fully explore: the way India circumvented the standard etiquette of the Antarctic Treaty System also circumvents the scrutiny of environmental impact assessments.

5. These members did not all join at once. Australia and Russia are the oldest members and Romania and China joined later. The member nations have varying degrees of influence and stature in this team.

6. Some of the discussions that took place at this time are not part of the record. Thus, I do not use identifying names or identify speakers by nation unless their contributions appeared in the public record, either in the ATCM reports or in the press.

7. One delegate stated that the ASMA management plan had been in draft since 1996, another said since 1999.

8. These are the SCAR/COMNAP meetings: the Scientific Committee on Antarctic Research holds meetings to showcase scientific research and to organize international research programs and strategies. The Council of Managers for National Antarctic Programs meets to report on and coordinate logistics that support Antarctic research operations.

9. It must be noted that the South American states also maintain a quite formidable bloc of power in the Antarctic Treaty System and usually manage to avoid the type of criticism leveled against India.

10. During another break, I spoke with a delegate about getting permission to sit in on the Larsemann Hills ASMA management team meetings at Hobart, Australia. (He was not able to get access for me.) This delegate took the opportunity to explain to me that because the Indians are not working in their first language and they don't have translators (Hindi is not an official Antarctic Treaty language), the members of the Indian delegation do not understand the process. (English is an official language of India.)

7. Charismatic Data and Climate Change

1. In New Zealand, this event was referred to as the Boxing Day tsunami. There is disagreement about which event in another part of the world caused the breakup because it occurred during winter, when there were few people in Antarctica and no one directly observed the exact date of the breakup.

2. At this mainstream, policy-oriented conference, "workable solutions" consisted of technocratic fixes, such as carbon taxes and cap-and-trade systems, or penalties for violating particular agreed-upon CO_2 limits.

3. Keeling's initial research was funded by the director of Scripps Institute at that time, Roger Revelle, an early figure in climate science and the person who introduced Al Gore to the idea of anthropogenic climate change.

4. See Howe (2014) for a more complete history of global warming that begins with the Keeling Curve.

5. This chapter does not interrogate the reality of climate change. Instead, I assume that the scientific consensus that is outlined, recapitulated, and renegotiated in exhaustive detail on a regular basis by the Intergovernmental Panel on Climate Change is correct. There are very few climate change skeptics among working research scientists and skeptics are not considered to be a scientifically credible minority opinion group because of their exceedingly small numbers.

6. While this chapter briefly discusses the 550 parts per million solution, it does not describe what the devastating ramifications would be for some people and nations if this CO_2 level was reached and the projected global temperature rise occurred. Such a solution almost certainly guarantees sea level rises that will flood many island nations and trigger a health, safety, and refugee crisis of critical proportions. The 550 parts per million solution is not the only proposed target number for carbon dioxide emissions. Current emissions stand at about 407 ppm and rising. At the Wellington Climate Change and Governance Conference, when policy was discussed, the numerical figure presented was always 550 ppm. It was the accepted policy solution number at this conference. A few speakers, in particular representatives of Pacific Island states, decried this "solution." For most, though, it seemed a workable goal for mitigating economic and environmental problems in the future associated with climate change. The 550 parts per million scenario is examined in detail by the Intergovernmental Panel on Climate Change (2007).

7. Bush eventually acknowledged the reality of anthropogenic climate change after the IPCC published its Fourth Assessment Report in 2007.

8. Antarctica New Zealand edits its official science strategy every five years or so, creating categories that scientists must use when they apply for funding. The first science strategy was written with input from Antarctic scientists at a workshop. More recent incarnations have been edited through e-mail participation.

9. CCAMLR is the other main international policy body relating to the Antarctic. It manages sustainable fisheries in the Southern Ocean based on annually collected data. CCAMLR meets yearly in Hobart, Tasmania, Australia. Many of the policy makers who attend the ATCM also attend the CCAMLR meetings.

REFERENCES

Abraham, Itty. 1998. *The Making of the Indian Atomic Bomb: Science, Secrecy, and the Postcolonial State*. London: Zed Books.

Adam, David. 2004. "The Iceberg Cometh." *The Guardian*, December 15. http://www.theguardian.com/world/2004/dec/15/research.antarctica.

Age, The. 2006. "Black Ice." July 17. http://www.theage.com.au/news/in-depth/black-ice/2006/07/16/1152988408997.html.

Anderson, Warwick. 2002. "Introduction: Postcolonial Technoscience." *Social Studies of Science* 32 (5–6): 643–658.

Anghie, Antony. 2005. *Imperialism, Sovereignty, and the Making of International Law*. Oxford: Cambridge University Press.

Antarctic and Southern Ocean Coalition. 2004. "Report of the Antarctic and Southern Ocean Coalition." ATCM Information Paper, Report on the XXVII Antarctic Treaty Consultative Meeting. March. http://www.asoc.org/storage/documents/Meetings/ATCM/XXVII/asoc.ip.report-1.pdf.

Antarctic Treaty Consultative Meeting. 1991. "The Protocol on Environmental Protection to the Antarctic Treaty." Madrid, Spain. http://www.ats.aq/e/ep.htm.

——. 2004. "WP 001. Review of Draft McMurdo Dry Valleys Antarctic Specially Managed Area Management Plan: Report of the Intersessional Contact Group (NZ)." Buenos Aires: Secretariat of the Antarctic Treaty.

———. 2006. *Final Report of the Twenty-Ninth Antarctic Treaty Consultative Meeting, Edinburgh, United Kingdom, 12–23 June 2006*. Buenos Aires: Secretariat of the Antarctic Treaty.

Antarctic Treaty Consultative Parties. 1959. *The Antarctic Treaty*. Buenos Aires: Secretariat of the Antarctic Treaty.

Arrhenius, Svante. 1896. "On the Influence of Carbonic Acid in the Air upon the Temperature of the Ground." *Philosophical Magazine and Journal of Science* 41 (251): 237–276.

Bamber, J. L., and W. P. Aspinal. 2013. "An Expert Judgment Assessment of Future Sea Level Rise from the Ice Sheets." *Nature Climate Change* (January 6). http://www.nature.com/nclimate/journal/v3/n4/full/nclimate1778.html.

Barker, K. 2008. "Flexible Boundaries in Biosecurity: Accommodating Gorse in Aotearoa New Zealand." *Environment and Planning A* 40 (7): 1598.

Bastmeijer, Kees, and Ricardo Roura. 2008. "Environmental Impact Assessment in Antarctica." In *Theory and Practice of Transboundary Environmental Impact Assessment*, edited by Kees Bastmeijer and Timo Koivurova, 175–219. Leiden: Brill/Martinus Nijhoff.

BBC News. 2005. "Pile-Up as Berg Hits Antarctica." *BBC News*, April 19. http://news.bbc.co.uk/2/hi/science/nature/4460335.stm.

Beck, Peter J. 1986. *The International Politics of Antarctica*. London: Croom Helm.

———. 2006. "The United Nations and Antarctica, 2005: The End of the 'Question of Antarctica'?" *Polar Record* 42 (3): 217–227.

Beck, Ulrich. 1992. *Risk Society: Towards a New Modernity*. London: Sage Publications.

———. 1995. *Ecological Politics in an Age of Risk*. Cambridge: Polity Press.

Bennett, Jane, and William Chaloupke. 1993. *In the Nature of Things: Language, Politics, and the Environment*. Minneapolis: University of Minnesota Press.

Bloom, Lisa. 1993. *Gender on Ice: American Ideologies of Polar Expeditions*. Minneapolis: University of Minnesota Press.

Bloomfield, Adrian. 2007. "Russia Claims North Pole with Arctic Flag Stunt." *The Telegraph*, August 1.http://www.telegraph.co.uk/news/worldnews/1559165/Russia-claims-North-Pole-with-Arctic-flag-stunt.html.

Boyer, Dominic. 2005. "The Corporeality of Expertise." *Ethnos* 70 (2): 243–266.

Brenneis, Donald. 1999. "New Lexicon, Old Language: Negotiating the 'Global' at the National Science Foundation." In *Critical Anthropology Now: Unexpected Contexts, Shifting Constituencies, Changing Agendas*, edited by George E. Marcus, 123–146. Santa Fe: School of American Research.

Brooks, Harvey. 1964. "The Scientific Advisor." In *Scientists and National Policy Making*, edited by R. Gilpin and C. Wright, 72–96. New York: Columbia Univ. Press.

Brosius, J. P. 1997. "Endangered Forest, Endangered People: Environmentalist Representations of Indigenous Knowledge." *Human Ecology* 25 (1): 47–69.

Brown, Paul. 1991. *The Last Wilderness: Eighty Days in Antarctica*. London: Hutchinson.

Bruner, Edward. 2004. *Culture on Tour: Ethnographies of Travel*. Chicago: University of Chicago Press.

Brysse, K., N. Oreskes, J. O'Reilly, and M. Oppenheimer. 2012. "Climate Change Predictions: Erring on the Side of Least Drama?" *Global Environmental Change* 23 (1): 327–337.

Burnett, D. Graham. 2000. *Masters of All They Surveyed: Exploration, Geography, and a British El Dorado*. Chicago: University of Chicago Press.

———. 2012. *The Sounding of the Whale: Science and Cetaceans in the Twentieth Century*. Chicago: University of Chicago Press.

Burroughs, Edgar Rice. 1922. *At the Earth's Core*. Chicago: A. C. McClurg.

———. 1923. *Pellucidar*. Chicago: A. C. McClurg.

———. 1930a. *Tanar of Pellucidar*. New York: Metropolitan.

———. 1930b. *Tarzan at the Earth's Core*. New York: Metropolitan.

———. 1937. *Back to the Stone Age*. Tarzana, CA: Edgar Rice Burroughs.

———. 1944. *Land of Terror*. Tarzana, CA: Edgar Rice Burroughs.

———. 1963. *Savage Pellucidar*. New York: Canaveral Press.

Byrd, Robert. 1930. *Little America: Aerial Exploration in the Antarctic, The Flight to the South Pole*. New York: G. P. Putman's Sons.

Callon, Michel, John Law, and Arie Rip. 1986. *Mapping the Dynamics of Science and Technology: Sociology of Science in the Real World*. Basingstoke, Hampshire: Macmillan.

Campbell, John W., Jr. as Don A. Stewart. 1938. "Who Goes There?" *Astounding Science Fiction* (August): 60–97.

Carey, Mark. 2010. *In the Shadow of Melting Glaciers: Climate Change and Andean Society*. New York: Oxford University Press.

Carr, E. Summerson. 2010. "Enactments of Expertise." *Annual Review of Anthropology* 39 (1): 17–32.

Carter, Paul. 1987. *The Road to Botany Bay: An Exploration of Landscape and History*. Chicago: University of Chicago Press.

Chait, Johnathan. 2014. "Why Do Republicans Always Say 'I'm Not a Scientist'?" *New York Magazine*. May 30. http://nymag.com/daily/intelligencer/2014/05/why-republicans-always-say-im-not-a-scientist.html.

Chandrakirti Meditation Centre. 2005. "Ven. Kirti Tsenshab Rinpoche Visiting Antarctica in Jan 2006 to Pacify World Weather Patterns." Accessed April 20, 2008. http://www.chandrakirti.co.nz/antarctica.php.

Chayes, Abram. 1996. *The New Sovereignty: Compliance with International Regulatory Agreements*. Cambridge, MA: Harvard University Press.

Cherry-Garrard, Apsley. 1929. *The Worst Journey in the World: Antarctic 1910–1913*. London: Constable and Co.

Chown, S. L., and P. Convey. 2007. "Spatial and Temporal Variability across Life's Hierarchies in the Terrestrial Antarctic." *Philosophical Transactions of the Royal Society of London, Series B* 362 (1488): 2307–2331.

Choy, Tim. 2011. *Ecologies of Comparison: An Ethnography of Endangerment in Hong Kong*. Durham, NC: Duke University Press.

Collier, Stephen J., and Andrew Lakoff. 2006. "Vital Systems Security." Discussion paper from the Laboratory for the Anthropology of the Contemporary. http://imap.www.anthropos-lab.net/wp/publications/2007/08/workingpaperno2.pdf.

Cook, Frederic Albert. 1900. *Through the First Antarctic Night, 1898–1899: A Narrative of the Voyage of the "Belgica" among Newly Discovered Lands and over an Unknown Sea about the South Pole*. New York: Doubleday and McClure.

Cook, James. 1961. *The Journals of Captain James Cook.* Vol. 2, *The Voyage of the Resolution and Adventure, 1772–1775.* Edited by J. C. Beaglehole. Cambridge: Cambridge University Press and Halykut Society.

Cruikshank, Julie. 2001. "Glaciers and Climate Change: Perspectives from Oral Tradition." *Arctic* 54 (4): 377–393.

———. 2005. *Do Glaciers Listen?: Local Knowledge, Colonial Encounters, and Social Imagination.* Seattle: University of Washington Press.

De Broyer C., P. Koubbi, H. J. Griffiths, B. Raymond, C. d'Udekem d'Acoz, A. P. Van de Putte, B. Danis, B. David, S. Grant, J. Gutt, C. Held, G. Hosie, F. Huettmann, A. Post, and Y. Ropert-Coudert. 2014. *Biogeographic Atlas of the Southern Ocean.* Cambridge: Scientific Committee on Antarctic Research.

Douglas, Mary, and Aaron Wildavsky. 1982. *Risk and Culture: An Essay on the Selection of Technological and Environmental Dangers.* Berkeley: University of California Press.

Dumit, Joseph. 2004. *Picturing Personhood: Brain Scans and Biomedical Identity.* Princeton, NJ: Princeton University Press.

Edwards, P. N. 2010. *A Vast Machine: Computer Models, Climate Data, and the Politics of Global Warming.* Cambridge, MA: MIT Press.

Eime, Roderick. 2005. "Outta the Way—It's B15A!" *World Adventurer*, February 5. Accessed March 29, 2016. http://worldadventurer.net/issue1/b15a.html.

Ferguson, James. 1990. *The Anti-Politics Machine: Development, Depoliticization, and Bureaucratic Power in Lesotho.* Cambridge: Cambridge University Press.

Fish, S. E. 1980. *Is There a Text in This Class? The Authority of Interpretive Communities.* Cambridge, MA: Harvard University Press.

Fleck, Ludwik. (1935) 1979. *Genesis and Development of a Scientific Fact.* Chicago: University of Chicago Press.

Fleming, Fergus. 2001. *Ninety Degrees North: The Quest for the North Pole.* New York: Grove Press.

Foucault, Michel. 1970. *The Order of Things: An Archeology of Human Sciences.* New York: Pantheon Books.

———. 1977. *Discipline and Punish: The Birth of the Prison.* New York: Pantheon Books.

———. 1978. *The History of Sexuality.* Vol. 1, *An Introduction.* New York: Vintage.

Fuchs, Sir Vivian. 1982. *Of Ice and Men: The Story of the British Antarctic Survey, 1943–1973.* Oswestry, Shropshire: A. Nelson.

Galison, Peter. 1997. *Image and Logic: A Material Culture of Microphysics.* Chicago: University of Chicago Press.

Galison, Peter, and Barton Bernstein. 1989. "In Any Light: Scientists and the Decision to Build the Superbomb, 1952–1954." *Historical Studies in the Physical and Biological Sciences.* 19 (2): 267–347.

Gieryn, Thomas. 1995. "The Boundaries of Science." *Handbook of Science and Technology Studies*, edited by Sheila Jasanoff, Gerald E. Markle, James C. Peterson, and Trevor Pinch, 393–443. Thousand Oaks, CA: Sage Publications.

Gitelman, Lisa. 2013. *"Raw Data" Is an Oxymoron.* Cambridge: The MIT Press.

Goodwin, C. 1994. "Professional Vision." *American Anthropologist* 96 (3): 606–633.

Grandin, Temple, and Catherine Johnson. 2004. *Animals in Translation: Using the Mysteries of Autism to Decode Animal Behavior.* Harcourt Press.

Green, Bill, and Craig Potten. 2003. *Improbable Eden: The Dry Valleys of Antarctica*. New Zealand: Craig Potten Publishing.

Greenhalgh, S. 2008. *Just One Child: Science and Policy in Deng's China*. Berkeley: University of California Press.

Gupta, Akhil. 1992. "The Song of the Nonaligned World: Transnational Identities and the Reinscription of Space in Late Capitalism." *Cultural Anthropology* 7 (1): 63–79.

Gurney, Alan. 1997. *Below the Convergence: Voyages toward Antarctica, 1699–1839*. New York: Penguin Books.

Gusterson, Hugh. 1996. *Nuclear Rites: A Weapons Laboratory at the End of the Cold War*. Berkeley: University of California Press.

Guston, David H. 1999. "Stabilizing the Boundary between US Politics and Science: The Role of the Office of Technology Transfer as a Boundary Organization." *Social Studies of Science* 29 (1): 87–111.

Guston, D. H., D. Sarewitz. 2002. "Real-Time Technology Assessment." *Technology in Society* 24 (1): 93–109.

Haas, Peter M. 1990. *Saving the Mediterranean: The Politics of International Environmental Cooperation*. New York: Columbia University Press.

Hamzah, B.A. 2010. "Malaysia and the Southern Ocean: Revisiting the Question of Antarctica." *Ocean Development & International Law* 41 (2): 186–195.

Haraway, Donna. 1989. *Primate Visions: Gender, Race, and Nature in the World of Modern Science*. New York: Routledge.

———. 1997. *Modest_Witness@Second_Millennium: FemaleMan©_Meets_OncoMouse™*. New York: Routledge.

———. 2007. *When Species Meet*. Minneapolis: University of Minnesota Press.

Harding, Susan. 2009. "Get Religion." In *The Insecure American: How We Got Here and What We Should Do About It*, edited by Hugh Gusterson and Catherine Besteman, 345–361. Berkeley: University of California Press.

Harris, Richard. 2006. "Alaskan Storm Plays Role of Butterfly for Antarctica." *All Things Considered*, October 5. National Public Radio. http://www.npr.org/templates/story/story.php?storyId=6204027.

Helmreich, Stefan. 2005. "How Scientists Think, about 'Natives' for Example: A Problem of Taxonomy among Biologists of Alien Species in Hawaii." *Journal of the Royal Anthropological Institute* 11 (1): 107–128.

———. 2009. *Alien Ocean: Anthropological Voyages in Microbial Seas*. Berkeley: University of California Press.

Henke, Christopher R., and Thomas F. Gieryn. 2007. "Sites of Scientific Practice: The Enduring Importance of Place." In *Handbook of Science and Technology Studies*, edited by Edward J. Hackett, Olga Amsterdamska, and Michael Lynch. 353–376. Cambridge: MIT Press.

Herzfeld, Michael. 1992. *The Social Production of Indifference: Exploring the Symbolic Roots of Western Bureaucracy*. Chicago: University of Chicago Press.

Herzig, Rebecca. 2005. *Suffering for Science: Reason and Sacrifice in Modern America*. Rutgers University Press.

Hetherington, Kregg. 2011. *Guerrilla Auditors: The Politics of Transparency in Neoliberal Paraguay*. Durham, NC: Duke University Press.

Heyman, Joseph McC. 2004. "The Anthropology of Power-Wielding Bureaucracies." *Human Organization* 63 (4): 487–500.

Holden, Constance. 1994. "Greening of the Antarctic Peninsula." *Science* 266 (5182): 35.

Howe, Joshua P. 2014. *Behind the Curve: Science and the Politics of Global Warming*. Seattle: University of Washington Press.

Hull, Matthew S. 2012. *Government of Paper: The Materiality of Bureaucracy in Urban Pakistan*. University of California Press.

Hustak, Carla, and Natasha Myers. 2012. "Involutionary Momentum: Affective Ecologies and the Sciences of Plant/Insect Encounters." *differences* 23 (3): 74-118.

Ingold, Tim. 2000. *The Perception of the Environment: essays on livelihood, dwelling, and skill*. Psychology Press.

Intergovernmental Panel on Climate Change (IPCC). 2007. *Climate Change 2007: Synthesis Report*. Final part of the Fourth Assessment Report of the Intergovernmental Panel on Climate Change. https://www.ipcc.ch/pdf/assessment-report/ar4/syr/ar4_syr.pdf.

———. 2013. *Climate Change 2013: The Physical Science Basis*. Contribution of Working Group I to the Fifth Assessment Report of the Intergovernmental Panel on Climate Change. Cambridge: Cambridge University Press.

———. 2014. *Climate Change 2014: Synthesis Report*. Contribution of Working Groups I, II. and III to the Fifth Assessment Report of the Intergovernmental Panel on Climate Change. Geneva: IPCC.

Jasanoff, Sheila. 1990. *The Fifth Branch: Science Advisors as Policymakers*. Cambridge, MA: Harvard University Press.

———. 2004. *States of Knowledge: The Co-Production of Science and the Social Order*. London: Routledge.

———. 2005. *Designs on Nature: Science and Democracy in Europe and the United States*. Princeton, NJ: Princeton University Press.

Jasanoff, Sheila, and Marybeth Long Martello. 2004. *Earthly Politics: Local and Global in Environmental Governance*. Cambridge: MIT Press.

Jayaraman, K. S. 2007. "India Plans Third Antarctic Base." *Nature* 447 (9).

Johnson, Nicholas, and Eirik Sonneland. 2005. *Big Dead Place: Inside the Strange and Menacing World of Antarctica*. Los Angeles: Feral House.

Joughin, I., B. E. Smith, and B. Medley. 2014. Marine Ice Sheet Collapse Potentially Under Way for the Thwaites Glacier Basin, West Antarctica. *Science* 344: 735–738.

Joyner, Christopher. 1998. *Governing the Frozen Commons: The Antarctic Regime and Environmental Protection*. Columbia: University of South Carolina Press.

Kamat, Prakash. 2006a. "India to Build Research Base in Antarctica." *The Hindu*, July 9.

———. 2006b. "Priority for Third Research Base on Larsemann Hills." *The Hindu*, May 26.

Keller, Evelyn Fox. 1983. *A Feeling for the Organism: The Life and Work of Barbara McClintock*. San Francisco, CA: W. H. Freeman and Company.

Kennedy, R. 2008. "Mortgaged Futures: Trauma, Subjectivity, and the Legacies of Colonialism in Tsitsi Dangarembga's *The Book of Not*." *Studies in the Novel* 40 (1–2): 86–107.

Kirksey, Eben. 2014. *The Multispecies Salon*. Durham, NC: Duke University Press.

Kirksey, Eben, and Stefan Helmreich. 2010. "The Emergence of Multispecies Ethnography." *Cultural Anthropology* 25 (4): 545–576.

Knorr Cetina, Karin. 1999. *Epistemic Cultures: How the Sciences Make Knowledge*. Cambridge: Harvard University Press.

Kohler, Robert. 2002. *Landscapes and Labscapes: Exploring the Lab–Field Border in Biology*. Chicago: University of Chicago Press.

Kohn, Eduardo. 2013. *How Forests Think: Toward an Anthropology Beyond the Human*. Berkeley: University of California Press.

Lahsen, M. 2005. Seductive Simulations? Uncertainty Distribution around Climate Models. *Social Studies of Science* 35: 895–922.

Lakoff, Andrew. 2006. "From Disaster to Catastrophe: The Limits of Preparedness." Understanding Katrina: Perspectives from the Social Sciences, June 11. http://understandingkatrina.ssrc.org/Lakoff/.

Lambek, Michael. 1993. *Knowledge and Practice in Mayotte*. Toronto: University of Toronto Press.

Langlitz, N., and S. Helmreich. 2005. "Biosecurity: A Response to Helmreich." *Anthropology Today* 21 (2): 20–21.

Latour, Bruno. 1983. "Give Me a Laboratory and I Will Raise the World." In *Science Observed*, edited by Karin D. Knorr-Cetina and Michael Mulkay, 141–170. London: Sage Publications.

———. 1987. *Science in Action*. Cambridge: Harvard University Press.

———. 1999. *Pandora's Hope: Essays on the Reality of Science Studies*. Cambridge: Harvard University Press.

———. 2004. *Politics of Nature: How to Bring the Sciences into Democracy*. Cambridge: Harvard University Press.

———. 2013. *An Inquiry into Modes of Existence: An Anthropology of the Moderns*. Cambridge, MA: Harvard University Press.

MacFarlane, Robert. 2003. *Mountains of the Mind*. New York: Pantheon Books.

Mallikarjun, Y. 2003. "India Plans to Set Up Permanent Base Station in Antarctica." *The Hindu*, September 2. http://www.thehindu.com/2003/09/02/stories/2003090205230400.htm.

Martin, Emily. 1994. *Flexible Bodies: Tracking Immunity in American Culture—From the Days of Polio to the Age of AIDS*. Boston: Beacon Press.

Massey, Doreen. 1994. *Space, Place, and Gender*. Cambridge: Polity Press.

McCaughrean, Geraldine. 2007. *The White Darkness*. New York: HarperTeen.

McCloskey, Deirdre N. 1985. *The Rhetoric of Economics*. Madison: University of Wisconsin Press.

Mercer, John. 1978. "West Antarctic Ice Sheet and CO_2 Greenhouse Effect: A Threat of Disaster." *Nature* 271: 321–325.

Miller, Phillip, and Peter Hanns Reill. 1996. *Visions of Empire: Voyages, Botany, and Representations of Nature*. Cambridge: Cambridge University Press.

Mitchell, Timothy. 2002. *Rule of Experts*. Berkeley: University of California Press.

Mukerji, Chandra. 1990. *A Fragile Power*. Princeton, NJ: Princeton University Press.

Nandy, Ashis. 1988. "Introduction: Science as a Reason of State." In *Science, Hegemony, and Violence: A Requiem for Modernity*, edited by Ashis Nandy, 1–23. Tokyo and Delhi: The United Nations University and Oxford University Press.

National Research Council. 2007. *Exploration of Antarctic Subglacial Aquatic Environments: Environmental and Scientific Stewardship*. Washington, DC: The National Academies Press.

Oppenheimer, M., B. C. O'Neill, M. Webster, and S. Agrawala. 2007. "Climate Change: The Limits of Consensus." *Science* 317: 1505–1506.

O'Reilly, J. 2011. "Tectonic History and Gondwanan Geopolitics in the Larsemann Hills, Antarctica." *PoLAR: Political and Legal Anthropology Review* 34 (2): 214–232.

———. 2015. "Glacial Dramas: Typos, Projections, and Peer Review in the Intergovernmental Panel on Climate Change." In *Climate Cultures: Anthropological Perspectives on Climate Change*, edited by J. Barnes and M. Dove, 107–126. New Haven, CT: Yale University Press.

O'Reilly, J., N. Oreskes, and M. Oppenheimer. 2012. "The Rapid Disintegration of Projections: The West Antarctic Ice Sheet and the Intergovernmental Panel on Climate Change." *Social Studies of Science* 42: 709–731.

Oreskes, Naomi. 1999. *The Rejection of Continental Drift: Theory and Method in American Earth Science*. Oxford: Oxford University Press.

Oreskes, N., K. Shrader-Frechette, and K. Belitz. 1994. "Verification, Validation, and Confirmation of Numerical Models in the Earth Sciences." *Science* 263: 641–646.

Orlove, Ben. 2005. "Human Adaptation to Climate Change: A Review of Three Historical Cases and Some General Perspectives." *Environmental Science & Policy* 8 (6): 589–600.

Peterson, M. J. 1988. *Managing the Frozen South: The Creation and Evolution of the Antarctic Treaty System*. Berkeley: University of California Press.

Pfeffer, W., J. Harper, and S. O'Neel. 2008. Kinematic Constraints on Glacier Contributions to 21st-Century Sea Level Rise. *Science* 321: 1340–1342.

Philo, Chris, and Chris Wilbert. 2000. *Animal Spaces, Beastly Places: New Geographies of Human-Animal Relations*. London: Routledge.

Pinch, Trevor J., and Wiebe E. Bijker. 1984. "The Social Construction of Facts and Artefacts: Or How the Sociology of Science and the Sociology of Technology Might Benefit Each Other." *Social Studies of Science* 14 (3): 399–411.

Poe, Edgar Allen. 1838. *The Narrative of Arthur Gordon Pym of Nantucket*. New York: Harper and Brothers.

Pratt, Mary Louise. 1992. *Imperial Eyes: Travel Writing and Transculturation*. London: Routledge.

Quigg, Phillip W. 1983. *A Pole Apart: The Emerging Issue of Antarctica*. New York: New Press.

Rabinow, Paul. 1996. *Making PCR: A Story of Biotechnology*. Chicago: University of Chicago Press.

Raffles, Hugh. 2002. *In Amazonia: A Natural History*. Princeton, NJ: Princeton University Press.

Rahmstorf, S. 2007. "A Semi-Empirical Approach to Projecting Future Sea-Level Rise." *Science* 315: 368–370.

Ramachandran, Sudha. 2007. "India's Polar Ambitions Are Growing." *Asia Times*, May 18. http://www.atimes.com/atimes/South_Asia/IE18Df07.html.

Ramaswamy, Sumathi. 2004. *The Lost Land of Lemuria: Fabulous Geographies, Catastrophic Histories*. Berkeley: University of California Press.

Ramesh, Randeep. 2006. "Blow to Plan for Polar Conservation Zone as India Joins the Cold Rush." *The Guardian*, August 11. http://www.theguardian.com/environment/2006/aug/11/india.conservationandendangeredspecies.

Reed, William. 1906. *The Phantom of the Poles*. New York: Walter S. Rockey Company.

Rignot, E., J. Mouginot, M. Morlighem, H. Seroussi, and B. Schuchl. 2014. "Widespread, Rapid Grounding Line Retreat of Pine Island, Thwaites, Smith, and Kohler Glaciers, West Antarctica, from 1992 to 2011." *Geophysical Research Letters* 41: 3502–3509.

Riles, Annelise. 2000. *The Network Inside Out*. Ann Arbor: University of Michigan Press.

——. 2006. *Documents: Artifacts of Modern Knowledge*. Ann Arbor: University of Michigan Press.

Rincon, Paul. 2006. "'Warm' Species Invading Antarctic." *BBC News*, June 23. http://news.bbc.co.uk/2/hi/science/nature/5101790.stm.

Rosaldo, Renato. 1989. *Culture and Truth: The Remaking of Social Analysis*. Boston: Beacon Press.

Science Daily. 2006. "Antarctic Ice Sheet Losing Mass, Says University of Colorado Study." *ScienceDaily*, March 2. Accessed April 21, 2008. http://www.sciencedaily.com/releases/2006/03/060302180504.htm.

Scott, James C. 1998. *Seeing Like a State: How Certain Schemes to Improve the Human Condition Have Failed*. New Haven, CT: Yale University Press.

Sellars, Richard West. 1999. *Preserving Nature in the National Parks: A History*. New Haven, CT: Yale University Press.

Sellen, Abigail J., and Richard H. R. Harper. 2002. *The Myth of the Paperless Office*. Cambridge, MA: MIT Press.

Shapin, Steven. 1994. *The Social History of Truth: Civility and Science in Seventeenth-Century England*. Chicago: University of Chicago Press.

Shapin, Steven, and Simon Schaffer. 1989. *Leviathan and the Air-Pump: Hobbes, Boyle, and the Experimental Life*. Princeton, NJ: Princeton University Press.

Sharma, Ashok. 2007. "Icy Continent Beckons Indian Researchers." *Financial Express*, May 14. http://archive.financialexpress.com/news/icy-continent-beckons-indian-researchers/155431.

Shore, Cris, and Susan Wright. 1997. *Anthropology of Policy: Critical Perspectives on Governance and Power*. London: Routledge.

Skrydstrup, M. 2012. "Modeling Ice: A Field Diary of Anticipation on the Greenland Ice Sheet." In *The Social Life of Climate Change Models: Anticipating Nature*, edited by K. Hastrup and M. Skyrdstrup. New York: Routledge.

Smith, R. I. Lewis. 2005. "The Thermophilic Bryoflora of Deception Island: Unique Plant Communities as a Criterion for Designating an Antarctic Specially Protected Area." *Antarctic Science* 17 (1): 17–27.

Space Daily. 2005. "Get Ready for the Largest Demolition Derby on the Planet." Space Daily, January 6. http://www.spacedaily.com/news/icebergs-04c.html.

Star, Susan Leigh, and James R. Griesemer. 1989. "Institutional Ecology, 'Translations,' and Boundary Objects: Amateurs and Professionals in Berkeley's Museum of Vertebrate Zoology: 1907–39." *Social Studies of Science* 19 (3): 387–420.

Steinberg, Phillip, Jeremy Tasch, and Hannes Gerhardt. 2015. *Contesting the Arctic: Politics and Imaginaries in the Circumpolar North*. London: I. B. Tauris.

Strathern, Marilyn. 2000. *Audit Cultures: Anthropological Studies in Accountability, Ethics, and the Academy*. London: Routledge.

Suess, Eduard. 1904. *The Face of the Earth (Das Antlitz der Erde)*. Oxford: Clarendon Press.

Sydney Morning Herald. 2006. "India to Build Antarctic Research Base," July 9.

Templeton, Malcolm. 2000. *A Wise Adventure: New Zealand in Antarctica, 1920- 1960*. Wellington: Victoria University Press in association with the New Zealand Institute of International Affairs.

Thomas, Craig W. 1997. *Bureaucratic Landscapes: Interagency Cooperation and the Preservation of Biodiversity*. Cambridge, MA: MIT Press.

Thomas, Nicholas. 2003. *Discoveries: The Voyages of Captain Cook*. London: Allen Lane.

Traweek, Sharon. 1988. *Beamtimes and Lifetimes: The World of High Energy Physicists*. Cambridge, MA: Harvard University Press.

Tsing, Anna. 2005. *Friction: An Ethnography of Global Connection*. Princeton, NJ: Princeton University Press.

Tufte, Edward R. 2006. *Beautiful Evidence*. Cheshire, CT: Graphics Press.

TV New Zealand. 2005. "Huge Iceberg Heading for Collision." TV New Zealand, January 13.

Van der Sluijs, Jeroen, Josee Van Eijndhoven, Simon Shackley, and Brian Wynne. 1998. "Anchoring Devices in Science for Policy: The Case for Consensus around Climate Sensitivity." *Social Studies of Science* 28 (2): 291–323.

Verne, Jules. 1864. *A Journey to the Center of the Earth*. France: Pierre-Jules Hetzel.

Vicuna, Orego. 1983. *Antarctic Resources Policy: Scientific, Legal, and Political Issues*. Cambridge: Cambridge University Press.

Weber, Max. 1947. *The Theory of Social and Economics Organization*. New York: Oxford University Press.

———. 1968. *Max Weber on Charisma and Institution Building: Selected Papers*. Edited by S. N. Eisenstadt. Chicago: University of Chicago Press.

Wilson, Eric G. 2003. *The Spiritual History of Ice: Romanticism, Science, and the Imagination*. New York: Palgrave Macmillan.

Wise, M. Norton, ed. 1995. *The Values of Precision*. Princeton, NJ: Princeton University Press.

Wolch, Jennifer, and Jody Emel. 1998. *Animal Geographies: Place, Politics, and Identity in the Nature-Culture Borderlands*. London: Verso.

INDEX